THE

PROGRESSIVE FARMER:

A Scientific Treatise

ON

AGRICULTURAL CHEMISTRY,

THE

GEOLOGY OF AGRICULTURE;

ON

PLANTS, ANIMALS, MANURES, AND SOILS.

APPLIED TO

PRACTICAL AGRICULTURE.

BY J. A. NASH,

PRINCIPAL OF MOUNT PLEASANT INSTITUTE, INSTRUCTOR OF AGRICULTURE IN AMHERST COLLEGE, AND MEMBER OF THE MASSACHUSETTS BOARD OF AGRICULTURE.

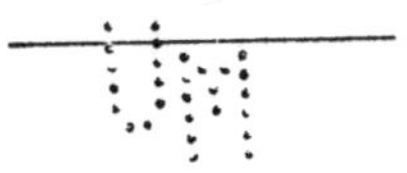

NEW YORK:
A. O. MOORE, AGRICULTURAL BOOK PUBLISHER,
(LATE C. M. SAXTON & CO.,)
No. 410 FULTON STREET.
1859.

Preface.

The undeniable fact, that some farmers are advancing in their profession, while others are retrograding, or only stationary, in connection with the author's belief, that *study* is the cause of success on one hand, and the want of it, of failure, on the other, will justify his choice of a name for this book —"The Progressive Farmer."

As Agriculture is necessarily a laborious employment—one in which a majority of mankind must ever be engaged, and on which all must depend for a subsistence—it is evident that whatever can be done to diminish its labors, to increase its profits, and to advance the intelligence and happiness of those who practise it, ought to be done.

The following pages are the result of an effort to render science available to practical farmers, to young men desirous of qualifying themselves for so useful an employment, and especially to the more advanced classes in our public schools.

With an earnest desire to contribute to the most important of all interests, and with a hope that the labor will not have been wholly in vain, these pages are submitted to the public.

J. A. N.

Contents.

CHAPTER I.

AGRICULTURAL CHEMISTRY.

CHAPTER II.

GEOLOGY OF AGRICULTURE.

CHAPTER III.

VEGETABLE PHYSIOLOGY.

CHAPTER IV.

ANIMALS AND THEIR PRODUCTS.

CHAPTER V.

MANURES.

CHAPTER VI.

PRACTICAL AGRICULTURE.

Introduction.

To "subdue the earth," to render it fruitful, and to keep it so, is the province of Agriculture.

Creative Power has made the earth capable of producing; has decreed that it *shall* produce *something;* but has left it for the skill and energy of man to decide, to a considerable extent, *what it shall produce*, and to determine, in some degree, *how much.*

In the first place, *the earth is to be subdued*, cleared of obstructions, mellowed, and cured of its tendency to useless production. In the second place, *useful productions are to be installed;* and these are to be selected with an intelligent reference to soil, climate, and the wants of the community. In the third place, *these productions are to be expended with a wise regard to future productiveness.* Such of their ingredients as came from the soil are to be returned to it, or others of equal fertilizing value to be substituted, in order that the soil may be increasingly fertile.

How best to prepare the soil—how to put it to the most profitable use—how to dispose of its products advantageously to both the soil and its owner, so that while the one shall increase in fertility, the other shall advance in wealth and intelligence, and in moral and social influence, are the questions of scientific agriculture.

Labor is an important requisite, but not the only requisite of successful husbandry. *Cultivated mind*, matured judgment, good sense enlightened by study and experience, find no better field on which to exert themselves than the farm. It cannot, indeed, be expected that practical men will acquire a profound knowledge of all the sciences which throw light on their path, for these are many and extensive.

Chemistry has made immense strides, and has achieved the most important discoveries. These must be brought to bear in favor of agriculture. *Geology,* though of recent origin, has already become a great and useful science. *Vegetable physiology* is replete with instruction to the farmer. *The history of animals* affords an almost limitless field of instruction. Because those who are, and those who intend to be practical farmers, cannot compass the whole of these and other sciences, it does not follow that they should cull nothing from them.

There are facts, principles, and conclusions from all the natural sciences, which can be easily acquired, and which cannot fail to be of the greatest service to practical agriculture. To state these facts, to illustrate principles, and to apply conclusions to the every-day business of the farmer, is the design of the following pages.

Should the first chapter appear to any too *difficult* and not sufficiently *practical,* I readily admit that it is difficult; it is so from the very nature of the subject, but it is not impractical. The subject of this chapter has important bearings on every branch of practical agriculture.

Succeeding chapters will be found more *directly* and *manifestly* practical—will have more and more to do, as we go on, with the every-day business of agriculture, and it is hoped, will become increasingly interesting and useful to *practical men.*

If *farmers* will peruse this and similar works, and will encourage their *sons* to *study* them, they will find that "it pays," both in the increased *pleasure* and in the augmented *profits* of agriculture.

CHAPTER I.

AGRICULTURAL CHEMISTRY.

EXPLANATION OF TERMS.

1. A BODY, that is constituted of one kind of matter only, is called an *element.*

2. One that is composed of two elements, is a *compound*, and is sometimes called a *binary compound*, to distinguish it from compounds containing more than two elements.

3. If a body consist of three elements, it is called a *ternary compound;* if of four, a *quaternary compound. Binary* implies two-fold; *ternary*, three-fold, and *quaternary*, four-fold.

4. Thus, iron being constituted of but one kind of matter, is an *element;* water being composed of two, is a *binary compound;* epsom salt, composed of three, is a *ternary compound;* and alum, of four, is a *quaternary compound.*

5. There are three forms in which bodies may exist —the *gaseous*, the *liquid*, and the *solid*. A body that is elastic, like air, is called a *gas;* one that is inelastic, like water, a *liquid;* and one in which the particles do not readily move among each other, as iron, wood, straw, feathers, a *solid*.

6. Some bodies are capable of assuming all these forms at different temperatures, as water, for instance, is gaseous above 212°, liquid from that down to 32°, and solid below that point.

7. Bodies which will combine with each other when brought into contact, are said to have an *affinity* for each other; those which will not, are said to have no such affinity. *Chemical affinity* is a tendency existing between certain bodies to combine and form compounds. It is of three kinds—*simple, single elective*, and *double elective*—*simple*, when two substances combine, no other body being present, as oxygen and hydrogen, to form water; *single elective*, when one substance decomposes another to combine with one of its ingredients, as when vinegar decomposes chalk, combining with its lime, and setting its acid free; and *double elective*, when two compounds exchange partners with each other.

8. We must distinguish between a *compound* and a *mixture*. When two substances combine of their own accord, as if self-moved, the result is a *compound*. If they are only put together by mechanical force, it is a *mixture*. In the first case, the properties of the ingre-

dients are entirely changed; in the last, they remain unaltered. Thus, if you bring chlorine and sodium together, a substance totally unlike either is produced; from two virulent poisons a wholesome condiment is formed—common salt: this is a *compound.* But if you put water with milk, no new substance is formed—the properties of the ingredients remain unaltered; they are water and milk still, and nothing more. This is a mere *mixture.*

9. A substance that can be dissolved in a liquid is said to be *soluble;* one that cannot, to be *insoluble,* as sugar, for instance, is soluble in water, and sand insoluble. When a substance is dissolved, it is called a solution, as a solution of sugar, salt, or nitre, in water. A distinction is also to be made between a *solution* and a *mixture.* If you put cider into water, this is nothing more than a *mixture;* a color is in this case communicated, whereas, if the cider were perfectly dissolved, it would leave the water transparent. If now you add a spoonful of salt to a pint of water, the water will remain as transparent as before. This is a *solution.* Any liquid which dissolves other substances is called a *solvent.* Water is the great solvent of those salts which feed growing plants. These salts enter the roots of plants in the state of transparent, colorless solutions in water.

10. There are different degrees of solubility. Water will hold in solution but $\frac{1}{600}$ of its own weight of quicklime; it will hold in solution $\frac{1}{500}$ of its own weight of gypsum; $\frac{4}{11}$ of its weight of common

salt; and much more of some other salts. Several substances are more soluble in cold water than in hot. Glauber's salt, for instance, is dissolved to a greater extent in cold, than in hot water. Common salt has the property of being equally soluble in cold water and in hot. If you put into 11 pounds of cold water 4 pounds of common salt, it will all be dissolved. If you put in more, all beyond 4 pounds will fall to the bottom undissolved. Precisely the same will take place if the water be hot. In either case the water will hold in solution 4 lbs. of the salt to 11 of its own weight. Most substances, as is well known, are dissolved more readily, and in larger amounts, in hot water than in cold.

11. It is a general law of chemical combination, that *elements will combine with elements only, and compounds only with compounds.* According to this law, a body that is constituted of one kind of matter only, will combine with another body similarly constituted, but not with one that is composed of two kinds; and a body, composed of two kinds of matter, will combine with another that is constituted similarly, but not with one that contains but one kind of matter.

12. *All chemical combinations are in certain, definite proportions.* Bodies will not combine in any proportions which the chemist might prefer, but only in certain proportions, fixed in nature, and unalterable. In illustration of these principles, it may be stated that 8 lbs. of oxygen, *an element*, will combine with 1 lb. of hydrogen, *another element*, and form 9 lbs. of water.

Also, calcium, an element, will combine with the element, oxygen, precisely 20 lbs. of the first to 8 lbs. of the last, and form 28 lbs. of quick-lime. Now, if we take these two compounds, water and quick-lime, 9 lbs. of the former will combine with 28 lbs. of the latter, and form 37 lbs. of slacked lime. It is true, you might put more than 9 lbs. of water to 28 lbs. of lime, but the excess would soon evaporate, leaving precisely 9 lbs. combined with the lime in the form of a dry, white powder, (water-slacked lime). If you were to put less than 9 lbs. of water to 28 of lime, then only a part of the lime would be slacked; and in order to slack the whole, you would have to continue putting on water till you had reached the 9 lbs., when the whole would be reduced to a dry, white powder. It is so with all chemical combinations; they are always in definite, fixed and unalterable proportions. In this respect they differ from mere mixtures, which may be in any proportions.

ELEMENTS.

13. There are in nature 15 simple substances, called elements, which make up more than 99 hundredths of all known matter. Other substances exist in small quantity, but these are all that need be noticed in an introduction to agricultural chemistry. They constitute essentially all the objects with which we are conversant. If we analyze a stone, a handful of earth, a plant, a flower, a bone, a drop of water, a piece of flesh, almost anything we can think of, it is found to consist of one, two, three or more of these; seldom of

one, oftener of two, very often of three, less frequently of four, and rarely of more than four.

14. The names of the 15 elements, mentioned above, as constituting more than 99 hundredths of all known matter, are 1. Oxygen; 2. Chlorine; 3. Sulphur; 4. Phosphorus; 5. Carbon; 6. Silicon; 7; Nitrogen; 8. Hydrogen; 9. Iron; 10. Manganese; 11. Potassium; 12. Sodium; 13. Calcium; 14. Magnesium; 15 Aluminum.

15. *Oxygen* is a gas, colorless, tasteless, inodorous; not distinguishable by any of the senses from common atmosphere. It constitutes, as mixed with nitrogen, 1-5 of the air; as combined with hydrogen, 8-9 of water; enters largely into all plants and animals; forms a part of rocks and soils; and is supposed to constitute not far from one half of all known matter. It is the great supporter of combustion; and it constitutes the respirable portion of the atmosphere. No fire can burn without it, nor animal breathe in its absence. It enters into combination with all other elements. We seldom see anything, unless it be the precious metals, which is not composed in part of this substance.

16. *Chlorine* is a yellowish green gas, $2\frac{1}{2}$ times heavier than air, existing largely in sea-water, constituting more than half of common salt, and entering in a slight degree into all soils, and forming a part of all plants. On soils found by analysis to be deficient in chlorine, it should be supplied in the form of common

salt; and when we are about to plant those crops, which require a large amount of chlorine, (corn, potatoes, turnips,) we should apply salt, unless pretty well assured that the soil is well supplied with it, especially at a great distance from the sea; for the risk of losing on a few bushels of salt, is less than that of losing on the crop for the want of it.

17. *Sulphur* is a yellow, solid substance, known as roll brimstone, flower of sulphur, and, in a still finer state, as milk of sulphur. It exists, in some parts of the world, as a considerable rock formation. It constitutes a part of all soils. The waters of many springs are impregnated with it. As certain portions of all plants and animals contain it in their composition, it must exist in the soil, from which these derive their nourishment.

18. *Phosphorus.*—A yellow, solid substance, of something like the consistency of bee's-wax, forming a part of the bones of all animals and of the seeds of many plants, diffused in small quantities through rocks and soils of the earth and through the waters of the ocean.

19. *Carbon.*—Diamond is pure carbon. Charcoal is pure carbon, with the exception of what remains as ash, after being burned. It exists in a gaseous state in the air, constituting about one part in six thousand of the entire atmosphere. Carbon forms a part of all plants and animals, and of nearly all minerals.

20. *Silicon* is the basis of sand, flint, and quartz. It enters largely into all soils, and constitutes probably about 1-5 of the solid globe. In its pure state it is a dark brown powder. Combined with oxygen, it forms the flinty stones so common everywhere; also sand, which is flint stone reduced to different degrees of fineness.

21. *Nitrogen.*—A gas, tasteless, colorless, inodorous, and a little lighter than common air. Mixed with oxygen, it constitutes 4-5 of the atmosphere. It enters into the composition of all animals, and of nearly all plants. It constitutes, with oxygen, nitric acid; and forms a part of all those salts called nitrates.

22. *Hydrogen* is a tasteless, colorless, inodorous gas, 14 times lighter than air, and used on this account for filling balloons. It constitutes 1-9 of water, and a part of all vegetable and animal substances. Oxygen is a supporter of combustion (causes other bodies to burn); Hydrogen is combustible (burns); Nitrogen is neither a supporter of combustion nor a combustible. Oxygen is also a supporter of respiration, as well as of combustion. Nitrogen is neither. No fire can burn nor animal breathe in it. And though Hydrogen burns, yet it is not a supporter of combustion. A burning body is extinguished if immersed in it.

23. *Iron.*—A well-known metal; cheap, because plenty; but, beyond doubt, the most useful of all metals.

24. *Manganese.*—A metal resembling iron, but of a darker color and more brittle. It is never found in its pure state; is prepared with great difficulty; and is in that state of no sort of use. It is found, combined with oxygen, in nearly all soils; and from the soil it enters into plants.

25. *Potassium.*—A brilliant, silver-white metal, with a high degree of metallic lustre; the metallic basis of potash; burns with great brilliancy if thrown upon cold water, or ice even; the lightest of all metals, being about 4-5 as heavy as water.

26. *Sodium.*—A white, silvery metal; the metallic basis of soda; burns if thrown upon warm water; 9-10 as heavy as water. Potassium and Sodium are the only metals known that are lighter than water.

27. *Calcium.*—A yellowish-white metal, the basis of lime. It is from calcium, the metallic basis of lime, that a limy soil is called calcareous.

28. *Magnesium.*—A white, shining metal, the basis of calcined Magnesia.

29. *Aluminum.*—A metal in the form of a gray powder; not easily melted; the metallic basis of clay and of clay soils.

30. Of these 15 elements, 4, when in an uncombined state, are gases, viz.: *Oxygen*, *chlorine*, *hydrogen*, and *nitrogen*. The remaining eleven are solids at ordinary temperatures.

31. Iron and manganese are *metals* proper, as distinguished from the alkaline and earthy metals.

32. Potassium and sodium are *metals of alkalies;* calcium and magnesium, of *alkaline earths;* and aluminum, of the *earth, alumina* (clay).

33. *Carbon, hydrogen, oxygen,* and *nitrogen* are called *organic* elements, because they constitute by far the larger part of all organized substances, whether animal or vegetable.

TABULAR VIEWS OF ELEMENTS, COMPOUNDS, AND SALTS.

34. The 15 elements, above described, will now be presented in tabular view, together with some of the more important compounds and salts derived from them. (See Table I.)

35. It will be noticed that there is a capital letter, or a capital and a small letter, placed after each element. These are called *symbols.* It is little else than a short-hand, and very convenient way of writing the words before them; as O, for Oxygen; Cl, for Chlorine; S, for Sulphur, &c. With three exceptions, these are the initials of the names. The exceptions are that K, stands for Potassium, Na, for Sodium, and Fe, for Iron. It is important that these symbols should be well fixed in the memory.

36. It will be seen also that after each symbol there

is a figure. These figures represent the *atomic* weight of all substances. All matter is believed to exist in *atoms*, or indivisible particles. The atom of hydrogen, which is the lightest of all bodies, is put down at 1. The atom of oxygen is known to be 8 times as heavy, and is therefore put down at 8; that of chlorine, for a like reason, at 36; of sulphur, 16; phosphorus, 32, &c. Now when elements combine with each other, they combine by atoms, one atom of one to one atom of another, two atoms of one to one atom of the other; and so on, either 1, 2, 3, 4, 5, 6, or 7 of one to one of the other; or, as sometimes happens, 3 of one to 2 of the other. This enables the chemist to tell beforehand precisely how much of one substance will combine with a given quantity of another. If you look at nitrogen in the table, you will perceive that the number against it is 14. Now if you wished to combine oxygen with 14 grs. of nitrogen, it would take just 8 grs., or just twice 8 grs., or three, four, five, six, seven times 8 grs. That is, oxygen will combine with nitrogen in the proportion of 8, 16, 24, 32, 40, 48, or 56 grs. of the former, to 14 grs. of the latter, but in no other proportions. So it is with all other substances; they combine in the proportions of their own atomic weight, as expressed by figures, or in the proportion of even times these numbers. This will be plainer as we proceed.

37. The compounds of oxygen with the elements arranged below it (so many of them as we shall notice in this work) are placed opposite those elements respectively. (See Table I.) Other compounds of

the elements with each other are arranged below; and the figures after each show from which two elements each comes; while the symbols will show, (when the learner becomes familiar with them), in what proportion the elements, in each case, enter into the compound.

TABLE I.

ELEMENTS.	OXYGEN COMPOUNDS.				
1. Oxygen, O, 8.					
2. Chlorine, Cl, 36.	1. Chloric acid, ClO^5, 76	from	1	and	2.
3. Sulphur, S, 16.	2. Sulphuric acid, SO^3, 40	"	1	"	3.
4. Phosphorus, P, 32.	3. Phosphoric acid, PO^5, 72	"	1	"	4.
5. Carbon, C, 6.	4. Carbonic acid, CO^2, 22	"	1	"	5.
6. Silicon, Si, 22.	5. Silicic acid, SiO^3, 46	"	1	"	6.
7. Nitrogen, N, 14.	6. Nitric acid, NO^5, 54	"	1	"	7.
8. Hydrogen, H, 1.	7. Water, HO, 9	"	1	"	8.
9. Iron, Fe, 28.	8. Oxides of Iron,*	"	1	"	9.
10. Manganese, Mn, 28.	9. Oxides of Manganese,	"	1	"	10.
11. Potassium, K, 39.	10. Potash, KO, 47	"	1	"	11.
12. Sodium, Na, 23.	11. Soda, NaO, 31	"	1	"	12.
13. Calcium, Ca, 20.	12. Lime, CaO, 28	"	1	"	13.
14. Magnesium, Mg, 12.	13. Magnesia, MgO, 20	"	1	"	14.
15. Aluminum, Al, 14.	14. Alumina, Al^2O^3, 52	"	1	"	15.

15. Chloride of Sodium, NaCl, 59	from	2	and	12.
16. Sulphuret of Iron, Fe^2S^3, 104	"	3	"	9.
17. Sulphuret of Hydrogen, HS, 17	"	3	"	8.
18. Light Carburet of Hydrogen, CH^2, 8	from	5	"	8.
19. Heavy Carburet of Hydrogen, C^2H^2,	"	5	"	8.
20. Ammonia, NH^3, 17	"	7	"	8.

* There are two oxides of iron, the protoxide and the sesquioxide. These are both important in their relations to agriculture, and will be explained fully in another place. There are also the protoxide and the peroxide of manganese.

TABLE II.

SALTS FORMED FROM THE FOREGOING COMPOUNDS.

1. Chlorate of Potash K O, Cl O^5, 123, from 1 and 10.
2. Sulphate of Iron (Copperas) Fe O, S O^3, 7 H O, 139, from 2 and 8.
3. Sulphate of Soda (Glauber Salt), Na O, S O^3, 10 H O, 161, from 2 and 11.
4. Sulphate of Lime (Gypsum, Plaster), Ca O, S O^3, 2 H O, 86, from 2 and 12.
5. Sulphate of Magnesia (Epsom Salt), Mg O, S O^3, 7 H O, 123, from 2 and 13.
6. Sulphate of Ammonia (soluble and fixed), from 2 and 20.
7. Phosphate of Lime (Bone Dust), about 2 parts lime to 3 of Phos. acid, from 3 and 12.
8. Super-phosphate of Lime, having more acid and less lime than the last, from 3 and 12.
9. Carbonate of Iron (Spathic Iron ore), Fe O, C O^2, 58, from 4 and 8.
10. Carbonate of Potash (Common Potash), K O, C O^2, H O, 78, from 4 and 10.
11. Bicarbonate of Potash (Saleratus), having twice as much acid as the last, from 4 and 10.
12. Carbonate of Soda (Washing Soda), Na O, C O^2, 10 H O, 143, from 4 and 11.
13. Bicarbonate of Soda (Cooking Soda), having twice as much acid as the last, from 4 and 11.
14. Carbonate of Lime (Chalk, Limestone), Ca O, C O^2, 50, from 4 and 12.
15. Carbonate of Ammonia (Volatile Ammonia in its most common form), from 4 and 20.
16. Silicates of Potash, Soda, Lime, Magnesia, &c. (in rocks and soils), from 5 and 8—14.
17. Nitrate of Potash (Nitre, Saltpetre), K O, N O^5, 101, from 6 and 10.
18. Nitrate of Soda (Soda-Saltpetre), Na O, N O^5, 85, from 6 and 11.
19. Nitrate of Lime (formed in limed muck-heaps and in old plaster), Ca O, N O^5, 82, from 6 and 12.
20. Chloride of Lime (bleaching, disinfecting, agricultural), composed of Chloric acid, Chlorine, and Lime.

38. In Table II. are arranged the principal salts (salts having special relation to agriculture), which are derived from the compounds, in the second column of Table I., and from other compounds at the bottom of that table. The figures placed after them show from which two compounds each salt is formed.

EXPLANATION OF THE FOREGOING TABLES.

39. Two things are essential to success in learning chemistry: 1st, to become able to infer from *the name* of a substance *what* it is composed of; and 2nd, to know how to name a compound from the names of its ingredients. You would suppose that if a chemist discovers a new compound, he may call it what he pleases. But it is not so; he must give it a name, which will indicate its ingredients, so that others may know, as soon as they hear its name, what it is made up of. Chemists have proceeded on this principle for the last half century; and it is due in no small degree to the excellence of their nomenclature, that they have achieved so many and so valuable discoveries. It is for the purpose of explaining the nomenclature of chemistry, that I have introduced the foregoing tables. The reader will notice that at the head of the table of oxygen compounds, we have six acids, each named after the element that combines with oxygen to form it; as sulphuric acid, from sulphur and oxygen; carbonic acid from carbon and oxygen; and so of the others. Besides these six acids there is another, which has intimate relations to agriculture, viz., *hydrochloric*

acid (H Cl), composed of one atom of chlorine, 36, to one of hydrogen, 1, making 37. In English works this last is usually called *spirit of salt*; in this country it is almost uniformly called *muriatic acid*, and will be so denominated in this work. We have then seven mineral acids; and the reader will perceive, if he looks at Table I., near the bottom of the oxygen compounds, that we have also 7 oxides, viz., oxide of iron, oxide of manganese, potash, soda, &c. Now, in order to form those combinations, commonly denominated salts, one of the foregoing seven acids must be combined with one of these oxides. From the fact, that the oxides constitute an important part of the salts, they are called also *bases*. For the purpose of aiding the memory, we will here arrange these acids and bases, together with the generic names of the salts, side by side.

TABLE III.

ACIDS.	BASES.	SALTS.
Chloric Acid,	Oxide of iron,	Chlorates,
Sulphuric Acid,	Oxide of Mn,	Sulphates,
Phosphoric Acid,	Potash,	Phosphates,
Carbonic Acid,	Soda,	Carbonates,
Silicic Acid,	Lime,	Silicates,
Nitric Acid,	Magnesia,	Nitrates,
Muriatic Acid,	Alumina,	Muriates.

40. There are other salts, formed in a different manner; as common salt, constituted of chlorine and sodi-

um, and some others; but the above, often called *oxygen salts*, as being composed in part of oxygen (except the muriates), are all formed from one of the above acids, and one of the accompanying bases. The name is decided, by changing the ending of the name of the acid, into *ate*, and then putting after it the name of the base, with *of* between. Thus, if we combine sulphuric acid with lime, it forms *sulphate of lime;* nitric acid with lime, forms *nitrate of lime;* carbonic acid with lime, *carbonate of lime;* carbonic acid with soda, *carbonate of soda;* silicic acid with potash, *silicate of potash;* and so of the others, each acid forming one or more salts with each base, and the salt in each case taking the names of both ingredients. When a second salt is formed from the same ingredients, it often takes a double portion of the acid, and then *bi* is put before the name, as a prefix. Thus, 22 parts, by weight, of carbonic acid with 31 parts of soda, form *carbonate of soda;* but 44 parts of carbonic acid to 31 of soda form bicarbonate of soda. The first is washing soda; the last, that kind of soda used in cooking. Sometimes the prefix, super, is used with the same meaning. You find the expressions *bicarbonate*, *supercarbonate*, *bisulphate*, *superphosphate*, and the like, all implying a double dose of the acid.

41. There is one thing that always troubles beginners in Chemistry: it is to distinguish between the substances whose names end in *uret*, and those whose endings are in *ate*. This difficulty should be conquered in the outset. Those substances whose names end in *uret*, are all the result of an *element* combined

with another *element;* those ending in *ate*, are in all cases the result of an *acid* combined with an *oxide*, or base. Thus, if you combine sulphur (an *element*) with iron (another *element*), you have a *sulphuret* of iron; but if you first combine sulphur and iron with oxygen, to form sulphuric acid and oxide of iron, and then combine these last with each other, you have a *sulphate* of iron. In other words, sulphur, phosphorus, and carbon, combined with any of the elements below them in Table I., form sulph*urets*, phosph*urets*, and carb*urets;* but if sulphuric acid, phosphoric acid, and carbonic acid combine with any of the bases below them in the second column of that table, they form sulphates, phosphates, and carbonates; and if twice the usual quantity of these acids are thus combined, they form *bi*sulphates, *bi*phosphates, and *bi*carbonates, as before explained.

42. If the learner is desirous of making real progress, he must master the principles laid down in the few preceding pages. This done thoroughly, he will find little difficulty. Let him turn back and review the brief description of the fifteen elements before given. Of these he needs to have as distinct, definite an idea as possible. Let him then look at Table I., and question himself on each of the binary compounds. On the first, he may inquire of what is chloric acid composed? The figures will point him to the two elements, and the symbol will show him in what proportion those elements combine to form it. The Cl shows him that chlorine is one of its elements, and the O shows him that oxygen is the other. The atom of

chlorine, he will see by casting an eye at the opposite column, is 36. In the same way he will see that the atom of oxygen is 8. But the small ^{5} after the O shows that there are 5 times 8 of oxygen to 36 of chlorine; that in 76 lbs. of chloric acid are 36 lbs. of chlorine and 40 lbs. of oxygen. If he look at the second compound, he will see that its symbol is $S O^3$, that is, sulphuric acid has one atom of sulphur, 16, and 3 of oxygen, 8 each, making 24; so that 40 lbs. of it would contain 16 lbs. of sulphur and 24 lbs. of oxygen. On coming to the eighth he will find no symbol. The reason is, that there are several oxides of iron, and they could not all be represented there. The two which have important relations to agriculture are the protoxide and the sesquioxide. It should be explained here that a *prot*oxide is one in which there is but one atom of oxygen to one of the metal; a peroxide, one in which there is much oxygen; and a sesquioxide, one in which there are three atoms of oxygen to two of the metal; that is, a protoxide implies a low degree of oxygen; a sesquioxide, a higher degree; a peroxide, a still higher degree; and an acid, a higher degree still. Accordingly, protoxide of iron ($Fe O$) implies one atom of iron, 28, to one of oxygen, 8; and sesquioxide ($Fe^2 O^3$) implies two atoms of iron, 28 each, to three of oxygen, 8 each. The same is true of manganese. There is the protoxide of manganese ($Mn O$), and the sesquioxide ($Mn^2 O^3$). The practical relations of these two metals, particularly of iron, will be shown in another place, and they will be seen to be very important to the farmer.

43 When the learner has been through with the compounds, and ascertained by their symbols and numbers how each one is composed, let him turn to Table II. and examine the salts in the same way. His mind will thus insensibly become familiar with the subject. Let him ask himself, on the first salt, of what two compounds is it made up? Let him trace it back to its two compounds, and then trace these compounds back to their elements. Then let him take the second in the same way. He will find that copperas contains 36 lbs. of protoxide of iron (Fe O) to 40 lbs. of sulphuric acid (S O^3), and that it consolidates in itself 63 lbs. of water (7 H O); that is, in 139 lbs. of this substance are 36 lbs. of protoxide of iron, 40 of sulphuric acid, and 63 of water. If he look at the third, he will find that sulphate of soda (Glauber's salt) is made up of soda (Na O), sulphuric acid (S O^3), and water (10 H O), 31 lbs. of the first to 40 of the second and 90 of the last, so that in 161 lbs. of the crystallized salts there are 90 lbs. of water. This, as in other similar cases, is called the *water of crystallization*. If this salt is exposed to the air, the water of crystallization passes off, and what was 161 lbs. of crystals becomes 71 lbs. of a white powder, but possesses equal value as before. The same is true of Epsom salt; the water of crystallization passes off, and leaves a white powder, much lighter than the crystals, but of equal value.

44. In the same way, if we take up sulphate of lime (plaster, gypsum), we find its symbol to be Ca O, S O^3, 2 H O. Ca O, implies one atom of lime, 28; SO^3,

one of sulphuric acid, 40; and 2 H O, two of water, 18; making 86. If this salt be heated to redness, the water of crystallization is driven off, and 86 lbs. of it become 68 lbs. Sixty-eight pounds of burnt gypsum are of equal value, therefore, with 86 lbs. of ground. The learner, it is presumed, can now go on, and analyze for himself the remaining expressions for salts in Table II., satisfying himself in each case, what are the ingredients of the salt; whether it contains in its crystallized state any water of crystallization; and, if any, how much. In this way he will learn the composition of many substances, and be rendering himself familiar with the language of chemistry.

The nature of these substances will next claim our attention. Occasional applications will be made to agriculture as we pass along; but such application will be reserved mainly for another part of this work.

COMPOUNDS.

45. Chloric acid (ClO^5, see Table II.) is a violent, powerful acid, having so strong affinity for all combustible substances, that it can hardly be preserved with safety.

46. *Sulphuric Acid* is a compound of great importance in the arts, and is beginning to be used extensively in agriculture. If it contained no water, we should have in 40 pounds of the acid 16 lbs. of sulphur and 24 lbs. of oxygen; but as it always contains water, more or less, these ingredients are of course less than 16 and 24 lbs. in 40, but are always in that

proportion to each other. Its purity is tested by its weight. The more water it contains, the lighter it is; and no one should buy it for good, unless it is once and ¾ as heavy as water. It has generally been retailed for 12½ cents a pound, but can now be procured for agricultural purposes at 2½ cents. It is a very powerful acid, and may undoubtedly be used to advantage in composting some manures, and especially for dissolving bones, to be used as fertilizers. It is more commonly known as *oil of vitriol.*

47. *Phosphoric Acid* (PO^5) exists largely in the bones of animals, and in the phosphate of lime, a mineral called *appatite*, and is found in all soils, not entirely exhausted by cropping. How best to restore it to soils deprived of it by bad management, so as to enable them to produce the cereals in abundance, will be considered in another place. It may be obtained in a pure state by burning phosphorus in oxygen gas. In this state it gathers moisture from the air, and assumes the appearance of a white, flaky cloud, but is readily absorbed by water, rendering it intensely sour.

48. *Carbonic Acid* (CO^2) is made up of 1 atom of carbon, 6, to 2 of oxygen, 16, making its atomic weight 22. That is to say, in 22 lbs. of carbonic acid are 6 lbs. of carbon and 16 of oxygen. This is a gas. It is 1½ times heavier than common air; and consequently, when produced in large quantities, it falls into low places, as dry wells, cellars, or cisterns, destroying sometimes the lives of those who descend; but, in accordance with a general law of gases, it soon

diffuses itself and mingles equally with the whole body of the atmosphere, forming on an average about 1-2500 of the whole. Water absorbs it in considerable quantity; and the more, if it is compressed, as in soda fonts. We know that plants are made up largely of carbon; in most cases not less than half their weight consisting of this substance. This carbon they obtain almost wholly from carbonic acid, which they receive by their leaves, from the air principally, but in a small part from the soil, as it enters their roots dissolved in water. The vegetation of the globe, therefore, is constantly abstracting immense amounts of carbonic acid from the air, enough to entirely deprive the whole atmosphere of this ingredient in a few years, if there were no re-supply. But when vegetable matters are burnt, when they are consumed by animals, and when they go to decay, their carbon is returned again to the air. If we eat a piece of bread, the carbon it contains combines with oxygen in the lungs, forming carbonic acid, and is thrown again into circulation in the atmosphere. So when wood, charcoal, pit-coal, tallow, oil, or any combustible matter, is burnt, the carbon they contain, and this is generally more than half of the whole, combines with oxygen and goes into the air, in the form of carbonic acid. Also when vegetable matter decays, the same thing happens. The process is slower, but the result is the same, so far as its carbon is concerned—that combines with oxygen by the slow process of decay, and goes again into general circulation, ready to be seized again by the leaves of plants, and again to be wrought into new vegetable forms. Lime-stone contains about 44 lbs. in one hundred of carbonic acid. When this is brought from the quarry and burnt into quick-lime, the carbonic acid is driven into the air. This is another source of re-supply. So when coal is drawn from the mine, and burnt, its carbon, long shut up in the bowels of the earth, is again set afloat for the use of plants. Many springs, as those at Saratoga, are throwing small, but constant streams of carbonic acid into the air. Volcanoes also, so long as active, are throwing out large quantities of it; and fissures in the earth, particularly in volcanic regions, often throw it out abundantly, and diffuse it through the atmosphere. It is true that large amounts of it are absorbed into the rivers, seas, and oceans, where it goes to support marine vegetation, to form the shells of fish, and to help build immense coral reefs; and some have feared that the atmosphere of the globe would ere long become so exhausted of it, as not to be able to sustain a vegetation equal to the growing wants of the race. But, when we consider the sources of re-supply above mentioned, we need not be alarmed; though it must be confessed that geology reveals a state of vegetation in by-gone periods, which proves that the atmosphere must have been more highly charged with this food of plants than at present. The fact that carbonic acid is a poisonous gas, and that it is always passing from the lungs of animals, shows the necessity of thorough ventilation in our rooms; and that our cattle even, though to be kept comfortably warm, should not be enclosed so tightly as to be compelled to breathe over their own breath.

Pure air, as we inhale it, contains about 1-2500 of this gas; as we exhale it, it contains 1-25, a hundred times as much as before; a very good reason, but only one among many, why we should not unnecessarily subject ourselves to the process of breathing the same atmosphere, over and over again—a good reason also, why the sexton should drive every particle of the old air out of the church between the morning and afternoon service, and why the teacher should ventilate thoroughly at noon and at the forenoon and afternoon recess, if not oftener.

49. *Silicic Acid* (SiO^3) is nothing else than sand, quartz, flint-stone, commonly caled Silica. A soil in which it abounds is called Silicious. It is composed of 1 atom of Silica, 22, to 3 of Oxygen, 24, forty-six pounds of it containing 22 lbs. Silicon, and 24 of Oxygen. It exists in the soil in two conditions, soluble and insoluble. When soluble, it is taken up by plants, and forms the stiffening of stems, straw, husks, &c. One office of manures, and especially of potash, soda and other alkalies, is to render a portion of the sand in the soil soluble, so that it may be available to plants. Soluble silica is essential to the perfection of most plants. Oats, grown on peat, for instance, will not mature straw sufficiently to support the grain. Nearly all soils, with the exception of peat, contain from 60 to 90 per cent. of silica.

50. *Nitric Acid* (NO^5) is composed of 14 lbs. of Nitrogen to 40 of Oxygen. It is a very powerful acid, known more commonly at the shops as aquafor-

tis. The salts formed by nitric acid are easily soluble. Hence they are uncommonly quick in their operation on plants. The Chinese gardener understands that by means of old plastering, which contains much nitrate of lime, he can force the growth of vegetables almost at pleasure, and cause an immense produce. In our country, such old plastering is too often thrown away.

51. *Muriatic Acid* (HCl) is 1 atom of Chlorine, to 1 of Hydrogen. Thirty-seven lbs. of it would give 36 lbs. of chlorine and one of hydrogen. It was formerly called *Spirits of Salt.* Its more appropriate name is Hydrochloric acid, because this name indicates the materials of which it is composed. But it is more commonly known in England as Spirit of Salt, and in this country as Muriatic acid.

52. *Water* (HO) is composed of 1 atom of Hydrogen, to 1 of Oxygen. Could you decompose a pint of water, it would give 1000 pints of Oxygen, and 2000 of Hydrogen. The Oxygen would weigh just 8 times as much as the Hydrogen, showing it to be just 16 times as heavy, by equal bulks. If now you should mix the two together, they would condense into 2000 pints; and if you then send an electric spark through them, they will combine into 1 pint of water. Consequently you perceive, that water must be just 1000 times heavier than Oxygen, and just 2000 times heavier than Hydrogen. Hydrogen is 16 times lighter than Oxygen and 14 times lighter than air, being, as before stated, the lightest of all known substances.

This is a well-known substance, and yet much is

to be learned of its various and vastly important offices in agriculture. We will enlarge on this subject at another time.

53. *Protoxide of Iron* (FeO) is a compound existing abundantly in many wet, marshy soils. It is largely soluble in water, and when so dissolved is injurious to vegetation, often preventing the growth of any thing save a little wiry, sour grass, which contains little or no nourishment. If such land be thoroughly drained, a large proportion of this oxide is taken off with the water; and what remains may be neutralized by ploughing and thus exposing it to the air; it takes another dose of oxygen, and becomes the red oxide or sesquioxide of iron, which is rather beneficial than hurtful to plants. The farmer may generally know whether his low lands are troubled with the protoxide of iron, by observing the water which flows from them. If impregnated with this oxide, it will generally show a film on its surface, often reflecting the colors of the rainbow. If this film be very thin, it reflects the yellow ray; if a little thicker, the red or brown; and if still thicker, the blue or violet. All these colors are sometimes reflected from neighboring points on the surface, which gives a sort of iris, or rainbow cast. The explanation is thus:—the protoxide of iron comes from the ground dissolved in water. On exposure to the air, it takes more oxygen and becomes the red or sesquioxide. This not being soluble in water, floats awhile on the surface, forming a film, varying in thickness, and, as before explained, in color, till at length it sinks to the bottom, giving the channel a sort of

yellowish-red appearance. Where these indications are presented, the land should be thoroughly drained in the first place; next, the soil should be turned up to the sun and air. Lime should then be applied if it can be obtained at a moderate price, say 20 or 25 cents a bushel; if not, ashes will do very well, but should by no means be applied till the land has become dry. Leached ashes for such a purpose, are worth probably somewhat more than half as much as unleached. If the ashes were to be applied before the water is removed, the leached would be just about as valuable as the unleached. Neither would be worth much. The potash and soda in the ashes would dissolve and run away with the water, and the lime, of which ashes contain some 75 per cent., would lie dormant in the soil.

54. *Sesquioxide of Iron* (Fe^2O^3) is composed of the same ingredients as the last, but contains, as the symbols show, a larger proportion of oxygen. The last, as before stated, changes into this, when exposed to the air. The scales and dust about the blacksmith's anvil are a mixture of those two oxides. These are a good dressing for fruit trees, but should be applied to the surface, instead of being dug in, in order that the black oxide may be exposed to the air, and thus have an opportunity of being converted into the red, or sesquioxide. It is this last oxide that gives to many soils their reddish brown color; and it is one or the other or both of these oxides of iron that give to so many subsoils their sickly yellow. Such subsoils are both *cold* and *poisonous* to plants; but they need only to be turn-

ed up to the sun and air, and properly manured, to become warm, healthy, and productive.

55. *Oxides of Manganese.*—These, like the oxides of iron, are numerous. Two—the protoxide (MnO) and the sesquioxide (Mn^2O^3)—are constituted similarly to the above oxides of iron. These are of little consequence to agriculture, and will not be spoken of again in this work. There is, however, another, which is of some importance to agriculture. It is the *peroxide*, or, as more commonly called, the *black oxide* of manganese (MnO^2), containing, as its symbol imports, one atom of manganese to two of oxygen. This exists in great abundance at Bennington, Vt., and at many other localities. It exists in small quantities in most rocks, and is slightly diffused through nearly all soils. It is found also in the ashes of most cultivated plants.

56. *Potash*, called by most writers potassa, (KO), is not the common potash of the shops, used for soap-boiling, but a far more bitter, acrid, caustic substance. It is seen at the apothecaries in the form of small, white rolls, not much larger than a pipe-stem, enclosed in vials air-tight, to prevent its taking carbonic acid from the air, and being turned to a carbonate of potash. Its caustic (burning) power is very great, so that it will readily dissolve horns, hoofs, bones, flesh, almost any animal matter. In order to form a correct idea of potash in all its changes, the learner must think first of a white, shining metal, like silver, so soft that you can cut it easily with a knife, and so light that it will float on water, almost instantly taking fire, and

burning brilliantly as it touches cold water or ice even. This is *potassium* (K). Now, if 8 parts, by weight, of oxygen be combined with 39 parts of this metallic potassium, we have *caustic potash* (KO), the intensely bitter, burning substance of which I have been speaking. If, then, 37 parts, by weight, of this caustic potash (KO) be combined with 22 of carbonic acid (CO^2), we shall have the common *carbonate of potash* of commerce (KO, CO^2). As found at stores, it is generally very impure. If now we take the dingy, gray potash of commerce, purify it of its foreign mixtures, and treat it to another dose of carbonic acid, we shall have saleratus, *bicarbonate of potash* (KO, $2CO^2$). Besides various other forms, we have then these four, in which potassium is exceedingly useful in the sciences, arts, and common affairs of life—viz., metallic *potassium* (K), caustic *potash* (KO), *carbonate of potash* (KO, CO^2), and *bicarbonate* (KO, $2CO^2$). In the form of common carbonate of potash only is it used for agricultural purposes. It is in this form that it exists in ashes. Ordinary wood ashes contain about 6 per cent. of carbonate of potash, some 2 per cent. of carbonate of soda, and about 75 per cent. of carbonate of lime. It is manifest, therefore, that farmers, who sell their ashes at the price generally paid by soap-boilers, and those who do not buy at these prices when they have an opportunity, commit a " mistake."

57. *Soda* (NaO).—This is caustic soda, consisting of sodium (Na), and oxygen (O). In this form it is useful in the arts and sciences, but is seldom seen or known in domestic concerns. Similar remarks apply

here as to potash. We have first metallic *sodium* (Na), a yellowish-white, shining metal, lighter than water, soft enough to be cut with a knife, that takes fire in warm water. Next, we have this metal, combined with oxygen only, soda, or oxide of sodium (NaO); then we have carbonate of soda, (washing soda), (NaO, CO^2); and then bicarbonate of soda, (cooking soda), ($NaO, 2CO^2$). I have not noticed the water (HO) in the foregoing combinations. The reader will perceive how much of it is consolidated in them by looking at Table II. It is in the form of carbonate of soda (NaO, CO^2), or ($NaO, CO^2, 10HO$), if we notice the water, that this substance is applied to soils. In this form it is used considerably in England, and is beginning to be used in this country. An impure kind of it is sold in the market as soda-ash. It is obtained from the ashes of sea-weeds.

58. *Lime* (CaO).—This is an oxide of calcium. It is lime as it comes from the kiln, before exposed to air. Lime in the quarry is the same substance, combined with carbonic acid. On being slacked it combines with water, 1 atom of water, 9, to 1 atom of lime, 28, making 37 for the atom of hydrate of lime. Thus 28 lbs. of quick-lime make 37 lbs. of dry slacked lime. Or if left after being taken from the kiln, exposed to the air, it first absorbs moisture, then crumbles to powder, and in a few days takes carbonic acid from the air, and becomes carbonate of lime (air slacked), just what it was in the quarry, except in structure. In tracing the metal, calcium, through some of its combinations, we have a course similar to those under pot-

ash and soda; first, we find metallic *calcium* (Ca); next, this, combined with oxygen only, *lime* (oxide of calcium), (CaO), (which is quick-lime, as it comes fresh from the kiln); we have also *carbonate of lime* (CaO, CO^2)—marble, limestone, chalk, and some varieties of marl; also the shells of insects and fish, are different forms of carbonate of lime, more or less impure. When lime combines with water, (consolidates water in itself, so as to be still apparently dry), it is called *hydrate of lime.* Such is the condition of water-slacked lime. Such also is the condition of many iron ores and other minerals. They consolidate in themselves large amounts of water, and yet are apparently dry. Such are called *hydrates*, as hydrate of lime, hydrate of iron, and others. From some hydrates the water is separated by a gentle heat; from others it cannot be driven off but by a very high heat.

59. *Magnesia* (MgO). This is the oxide of magnesium. It is known as calcined magnesia. Some impure lime-stones, as those called *dollomite* in Berkshire county, Mass., contain large quantities of carbonate of magnesia, in some cases not less than 40 per cent. This is often called magnesian lime-stone. If the carbonic acid be driven off by heat, a light, dry, white powder remains. This is calcined magnesia.

60. *Alumina* (Al^2O^3), as its name imports, is a compound of aluminum and oxygen, two atoms of the former to three of the latter. Alumina is a perfectly white powder, and is the basis of all clay soils. Pure

clay is a silicate of alumina, composed of about 60 per cent. of silica, and 40 of alumina.

61. *Chloride of Sodium* (NaCl) is composed of one atom of chlorine, 36, to one of sodium, 23. (See Table I.) It is no other than common salt. As corn, potatoes, and turnips contain large amounts of both its ingredients, it would seem hardly possible but that it should prove beneficial to these crops, especially on lands where either of them have been raised so long as to have exhausted the soil of the chlorine and sodium originally contained in it.

62. *Sulphuret of Iron* (Table II., 16).—There are three combinations of sulphuret and iron.

1st. The protosulphuret of iron (FeS), consisting of one atom of iron (Fe) to one of surphur (S).

2nd. The sesquisulphuret (Fe^2S^3), consisting of two atoms of iron to three of sulphur.

3rd. The bisulphuret (FeS^2), consisting of one atom of iron to two of sulphur. This last is often called fool's gold, from its strong resemblance to that metal.

63. *Sulphuret of Hydrogen*, or sulphuretted Hydrogen (HS), is a combination of one atom of sulphur, 16, to one of hydrogen, 1, making the atom of the compound 17. The nitrogenous, or azotized parts of plants and animals, contain a little sulphur and a very little phosphorus. When those substances which contain sulphur, as wool, hair, horns, hoofs, and eggs, de-

cay, it very often happens that an atom of the sulphur combines with one of hydrogen, and forms this gas. It may be recognized in the smell of rotten eggs, also about the docks in cities, and frequently in sinks. This gas is exceedingly unhealthy, as well as very oppressive, and it should never be tolerated about our buildings. The matter which gathers about the outlet of the sink should be frequently removed, or should be so diluted with peat or loam, with the addition of a little plaster or chloride of lime, as to give off no offensive odor, as this sulphuretted hydrogen is very apt to be generated in such places, and to operate injuriously on the health of families.

Sulphuretted hydrogen is formed in well-manured soils, and it is probably from this that plants obtain in part the sulphur, which they require in order perfectly to develop their seeds. It is a gas; but it readily dissolves in water; in which form (that of a limpid solution) it may enter the roots of plants.

64. *Carburet of Hydrogen* (CH^2 and C^2H^2) is of two kinds. (See Table I., 18.) Light carburetted hydrogen is composed, as its symbol imports, of carbon one atom, hydrogen two. This is the gas which often forms bubbles on the surface of stagnant water. It is inflammable. If you thrust down a pole into the bottom of water in which vegetable matter is decaying, bubbles will rise and float on the surface. These will burn with a gentle explosion and a whitish flame, if a torch be applied. This same gas is generated in richly manured soils, and probably it has something to do with furnishing plants with a small part of their food.

Heavy carburetted hydrogen (C^2H^2) is the gas used for lighting. It contains, as shown by its symbol, just twice as much carbon as the other, in consequence of which it gives a much stronger light. Heavy carburetted hydrogen may be obtained from almost any substance that contains carbon and hydrogen, as coal, oil, bark of trees, meats of nuts, &c., by heating it, with exclusion of air. If you put a walnut meat into the bowl of a tobacco pipe, cover it over with clay, and then thrust it into the fire, with the stem projecting upwards, this gas will soon issue from the stem. If you light it with a candle, you will have a good sample of a gas-light in a small way.

65. *Ammonia* (NH^3) is composed of one atom of nitrogen, 14, to 3 of hydrogen, 1 each, making 17. Consequently 17 lbs. of ammonia contain 14 lbs. of nitrogen and 3 of hydrogen. The peculiar odor of this compound may be recognized in the hartshorn of the shops, when used with quick-lime in the preparation of smelling bottles. It is generated whereever animal matter is undergoing decomposition; and if left to its own course it quickly combines with carbonic acid, forming a volatile carbonate of ammonia, and passes off into the air, to be blown about by the winds, and at length to be intercepted and brought back to the earth in the falling rains. In this way it is made to contribute as much to the growth of the useless as of the useful plants; for the rain, charged with this ammonia, falls as much on the wild mountain as on the cultivated plain. There are various easy and cheap modes of preventing its escape, which

will be explained in another part of this work, in connection with the use of fertilizers, the composting of manures, the husbanding of resources for the growth of plants, and other topics of practical agriculture.

A brief description has now been given of the 15 *elements*, which, in their various combinations, constitute nearly the whole of all known matter. (Table I., 1st column, 1–15.)

A very imperfect (because too short) account has been given also of 20 important *compounds* derived from those 15 elements. (Table II., 2nd column, and below 1–20.)

Of the *formation* of *salts*, by the combination of acids with bases (see Table III.), something has been said.

A consideration of the *nature* of salts, and of their use in agriculture, will be reserved for another place.

CHAPTER II.

GEOLOGY OF AGRICULTURE.

FORM OF THE EARTH---ITS DENSITY---PROPORTION OF LAND AND WATER---INEQUALITY OF SURFACE---WEIGHT OF ATMOSPHERE---CRUST OF THE EARTH.

66. THE earth has the form of an oblate spheroid, having an equatorial diameter 26 miles greater than its polar diameter. As this is the form, very nearly, which a fluid body would naturally assume, if revolving on its axis at the same rate, a fair inference is, that the earth was once in a fluid state. Its average weight is about 5 times that of water, and not far from twice and a half that of common rock.

67. About one fourth of the earth's surface is dry land, and three fourths are water. The land occupies not far from 50 million square miles, and the water about 150 million. The highest peaks of dry land are nearly six miles above tide water, and the lowest depths of the oceans are probably somewhat farther below. These inequalities affect the roundness of the

earth about as much as the smallest dust would that of an artificial globe. The average height of the land is probably a little less, and the average depth of the ocean a little more, than two miles.

68. The crust of the earth, thinner *comparatively*, there is reason to believe, than the shell of an egg, though certainly many miles in thickness, is solid rock, covered, three-fourths, as before stated, with water, and the remaining fourth, with broken rocks, stones, rounded pebbles, gravel, sand, and clay, to a depth of from a few inches to a few hundred feet; the whole sustaining an atmosphere supposed to be about 45 miles in height, and known to weigh just about 15 pounds to each square inch of the earth's surface. The weight of air over each square foot of the earth's surface is 2160 pounds; and the weight of the whole atmosphere is equal to the weight of a covering of water over the entire globe 34 feet deep. This is known from the action of a common suction pump, in which the pressure of the atmosphere just balances a column of water 34 feet high.

STRATIFIED AND UNSTRATIFIED ROCKS.

69. Almost every one must have noticed that some rocks, as they appear in various situations exposed to the eye, are formed into regular layers, or beds, resting one upon another. These layers are called *strata*, and the rocks that exhibit them are said to be stratified. Other rocks present no such appearance of *stratification*—no regular layers one upon another, and

are therefore said to be *unstratified.* The proof is very complete, though it cannot be given here, that the unstratified rocks were formed by fire, and that they took the form in which they appear by cooling off after being intensely heated. For this reason geologists have called them *igneous* rocks; and, because some portions of them have a crystallized appearance, they are often called *crystalline* rocks. We have then a class of rocks called indifferently *unstratified*, *igneous*, and sometimes *crystalline*, rocks, whose origin evidently was by fire.

70. It is, perhaps, equally well proved, and is, besides, a dictate of common sense, that the stratified rocks must have received their present form by deposition from water. For this reason they are often called *aqueous* rocks, and because most of them contain fossil remains of plants and animals, they are also called *fossiliferous* rocks.

71. If you were to see, on a steamboat, a row of huge casks, then above them a row of boxes, above these a row of bags, and above all, baskets, bundles, and umbrellas, you would have no hesitation in deciding which had been put there first. No one would dream that the pile had been commenced at the top and built downwards. The casks must have been rolled in first, the boxes placed on them, then the bags, and last of all the lighter matters. Equally clear are the reasonings of geologists. The lower rocks are older; and the higher are newer, with some exceptions, to be explained hereafter.

RELATIVE AGE OF ROCKS.

72. No man in his senses, and with any knowledge of the facts bearing on the question, would contend that the igneous and the aqueous rocks were formed at the same time. Either the heat, requisite to form the igneous rocks, would have expelled the water necessary to form the aqueous; or the water, necessary to form the aqueous, would have overcome the heat requisite to form the igneous. As well might you tell me that one piece of beef will bake and another freeze in the same oven and at the same time, or that the heat that will melt rocks will not convert water into steam. I would sooner believe either of these things than believe that the upper igneous and the lower aqueous rocks were formed at the same period. In the first place, it seems impossible that this could have been done, as much so as that the same oven could bake and freeze at the same time. In the second place, the aqueous rocks, with a few exceptions, easily accounted for, always lie above the igneous, showing thereby that they were deposited last. And in the third place, the aqueous rocks were manifestly formed out of the igneous, and therefore must have been formed subsequently. If a horse-shoe is made of iron, the iron must have been made first. Such are the reasonings of geologists with regard to the relative age of rocks, and those who doubt their main conclusions are generally those who have looked little at the facts.

CLASSIFICATION OF ROCKS.

73. As we come above the igneous, or unstratified, into the stratified rocks, we find them of many varieties, all of which have been arranged into three principal classes—*primary*, *secondary*, and *tertiary*.

74. The *primary* rocks either lie nearly horizontally upon the igneous, or lean with a gentle slope against them. In cases of the latter kind, it is believed, that the igneous rocks have been forced upwards by internal convulsions of the earth, and have raised the primary rocks along with them, inasmuch as all stratified rocks, having been deposited by the agency of water, must originally have been nearly horizontal. These primary rocks are generally hard. They have been subjected to immense pressure. Many of them bear marks of having been intensely heated since their deposition. Some of them are highly crystalline. They are nearly destitute of fossil remains, and the few they contain are entirely unlike any plants or animals now on the globe—an additional proof that, though not as old as the igneous rocks, on which they lie, they are older than other rocks which lie above them, and which contain fossil remains more like existing species.

75. Rocks of the *secondary* class overlay those of the primary; they contain more fossil remains; and the fossil remains found in them, though unlike existing species, bear a nearer resemblance to them than those in the primary rocks. These facts show them to be of later origin than the primary.

76. Rocks of the *tertiary* class are characterized by containing, among other fossil remains, species of animals, which are identical with those now on the earth. These overlay the secondary, and abound more than either of the others in fossil remains.

77. Over the tertiary rocks, and covering large portions of the earth, is what geologists have called *drift*—boulder rocks, rounded stones and pebbles, coarse and fine gravel, sand and clay, forming, in many cases, the soil which we now cultivate. This, all over the northern half of the globe, seems to have been transported, by some astonishing power, acting from the north, and carried in a southern direction, from a few rods to several hundred miles, from the rocks, in which it had its origin.

78. Since the drift period, various changes have taken place, and are still going on, as the result of causes now in operation, such as the running of streams, the filling up of ponds, and others. Strata, formed by these existing causes, are called *alluvial.*

79. We have then, as the most recently formed strata, *alluvial* deposits, next *drift*, next *tertiary* rocks, next *secondary*, and then *primary*, resting on the upper portion of the igneous rocks.

80. Among the igneous, or unstratified rocks, are granite, trap rocks, and the older and more recent lavas. These appear to have been ejected in a state of fusion by heat, at different epochs, from the bowels

of the earth, and to have consolidated, sometimes among the stratified rocks, and sometimes above them all, forming in some cases immense mountain masses of igneous rocks.

81. It is often said by those who have looked but little at this subject, that geologists know nothing about the comparative age of rocks; that God could have created the world at once, just as it is, with all its appearances of hoary age about it, with all its signs of ancient upheavings and volcanic vomitings, with its innumerable monsters imbedded within, creatures great and small, beautiful and ugly, formed as if for flying, running, swimming and creeping, but destined to do neither—all for no conceivable purpose, unless it were to deceive modern geologists. That God could do all this, I suppose no one wishes to deny. That He would do it, if there was a good reason for it, I have no doubt.

82. If I should say of an old book, dated a century ago, with as many dates scribbled on its margins, as there have been years since, with its binding well worn and its leaves thoroughly soiled, that there was no evidence of age about it—for the book-maker could manufacture just such a book as it now is—I should probably not be thought to reason very soundly; and yet the argument would be as good in one case as in the other, but for a single consideration, and that is, that a book-maker *can* deceive; God *will* not. To a reasoning mind there can be no doubt that the different portions of the earth's crust were formed at different and immensely distant periods.

ORIGIN OF SOILS.

83. All soils, whether alluvial, drift, or tertiary in their origin, are derived from rocks, broken down, ground to a greater or less degree of fineness, and so disseminated that the ruins of one rock may be supposed to be mixed, in most cases, with those of a great many others. The idea that soils have originated from the rock immediately under them is an error. When the drift period was, is not known, except that it was subsequent to the tertiary and anterior to the historic period; nor is it known what the drift agency was; but it is known, as well as anything can be, that some tremendous power was at work tearing up, transporting, and mixing the loose materials on the earth's surface. The soil on nearly every foot of land in our country—and the same is true of Europe, at least, if not of the whole world—has come from many and wide-spread localities. Every soil may be considered as a mixture of many soils. If every particle in a cubic foot of earth were to be endowed with instinct, and were to rise up and take its departure for its original rocky home, I have no doubt there would be a wide scattering, and I believe an extent of travel would be shown quite surprising to those who have not reflected on the subject.

84. If these views are correct—if the loose materials on the earth's surface have been extensively transported, scattered, and mixed—if they are now so mingled and confounded that the acutest geologist can detect the origin of only the coarser parts (boulders,

pebbles, and coarse gravel)—if, with regard to these, he finds the original locality from one to five hundred miles distant, all of which is sustained by the very best authorities, no one, that I know, disputing—it follows, of course, that soils depend very little, for their composition and capabilities, upon the rocks immediately underlaying them. This view is confirmed by analyses of soils. No more carbonate of lime, for instance, is found in lime-stone regions than in others. The same is true of other ingredients of soils. They are not always found in soils overlaying the rocks that contain them. Soils do not come from the underlaying rock, but from wide-spread regions, generally north and north-west of their present location. Hence, if rocks were ever so varied in their constitution, it would not follow that soils are.

ROCKS AND MINERALS.

85. The truth is, that rocks themselves are not as various in composition as many suppose. "Seven or eight simple minerals constitute the great mass of all known rocks. These are—1, quartz; 2, felspar; 3, mica; 4, hornblende and augite; 5, carbonate of lime; 6, talc, embracing chlorite and soap-stone; 7, serpentine. Oxide of iron is also very common, but does not usually show itself till the decomposition of the rock commences."—*Hitchcock's Geol., p.* 45.

86. From the same high authority we learn that "The following constitute nearly all the binary compounds of the accessible parts of the globe: 1, silica;

2, alumina; 3, lime; 4, magnesia; 5, potassa; 6, soda; 7, oxide of iron; 8, oxide of manganese; 9, water; 10, carbonic acid." It should be observed that every one of these binary compounds are formed out of the fifteen simple elements heretofore described, under the head of chemistry (Table I.). Every one of them is a compound of oxygen with one other element, so that only eleven of the elements enter into their composition.

87. Perhaps, for some of my readers, a description of the before-mentioned minerals may be needful.

Quartz is of various colors, but generally almost white; and when crystallized, it is transparent, a hard, flinty substance, composed almost wholly of silica, or silicic acid (SiO^3), known as flint, flinty stones and sand.

Felspar exists in connection with quartz and mica in granite, and may be distinguished from either by a glossy fracture when broken, somewhat resembling that of fine earthenware.

Mica, the third constituent of granite, is known extensively as isinglass, is of various colors, but more commonly nearly colorless, divisible into thin, flexible plates.

Hornblende is a common mineral, of various colors, occurring sometimes massive, at others in crystals; the crystals are sometimes short, but more generally long and slender, blade-like, sometimes fibrous.

Carbonate of Lime is a ternary compound, as its name implies; oxygen and calcium first uniting to form lime (oxide of calcium), and then carbonic acid uniting with lime to form the carbonate. It is known in various forms, as fine marble, common lime-stone, and chalk. It can be distinguished from almost any other mineral by its effervescence (bubbling), if an acid (vinegar, for instance) be poured upon it.

Talc is a magnesian mineral, consisting of broad, smooth laminæ, or plates. It is soapy to the touch; admits light through it; and is sometimes even transparent.

Chlorite and *Soap-stone* are little else than varieties of the same mineral.

Serpentine is also a magnesian mineral, of a greenish color, with spots resembling a serpent's skin—from which its name.

88. I have just quoted the opinion of a very eminent geologist, that these seven minerals "constitute the great mass of all known rocks," as also his opinion that silica, alumina, lime, magnesia, potash, soda, oxide of manganese, oxide of iron, water, and carbonic acid "constitute nearly all the binary compounds of the accessible parts of the globe." I will now invite attention to the opinions of the same writer with regard to the proportions in which these last-mentioned substances exist.

89. "It has been calculated that oxygen constitutes 50 per cent. of the ponderable matter of the globe; and that its crust contains 45 per cent. of silica, and at least 10 per cent. of alumina. Potassa constitutes nearly 7 per cent. of the unstratified rocks; and enters largely into the composition of some of the stratified class. Soda forms nearly 6 per cent. of some basalts.

and other less extensive unstratified rocks; and it enters largely into the composition of the ocean. Lime and magnesia are diffused almost universally among the rocks, in the form of silicates and carbonates—the carbonate having been estimated to form one-seventh of the crust of the globe. At least three per cent. of all known rocks are some binary combination of iron, such as an oxide, a sulphuret, a carburet, &c. Manganese is widely diffused, but forms much less than one per cent. of the mass of rocks."—(*Hitchcock's Geol., p.* 45.)

90. The foregoing is rather a geological than a chemical view. Most of the substances spoken of exist in rocks and soils, as ternary compounds. Says Dana (Muck Manual, p. 56—an unpretending name, but an excellent book), "Viewed in the light of chemistry, rocks are masses of silicates. The simple minerals composing rocks are truly only silicates in fixed proportions. The simple minerals are quartz, felspar, mica, hornblende, talc, serpentine."

91. According to this same author, chemically defining the above minerals, quartz is nearly pure silica; felspar and mica are silicates of alumina and potash; hornblende is silicate of alumina and lime, with magnesia; and talc and serpentine are silicates of magnesia. Thus it will be seen that silex, silica, or silicic acid, as unfortunately it is variously called, forms a very prominent part of the principal minerals, with the exception of carbonate of lime; and consequently of all rocks, except lime-stone; and then, as another conse-

quence, of all soils, inasmuch as soils are formed from rocks. If we consider that quartz, by far the most abundant mineral in nature, is nearly pure silica, and that the other leading minerals are more than half silica, we need not be surprised to learn that soils contain all the way from 60 to 90 per cent. of this ingredient. Sandy soils contain a higher per cent. still. Peats and bogs may be excepted, as not being strictly soils, but rather collections of organic matter—partially decayed vegetables. The average of silica in soils cannot be less than from 75 to 80 per cent.

92. From an inspection of analyses of rocks by distinguished chemists, it appears that the older rocks contain rather more silica, and a little less magnesia, alumina and lime, than the newer. If this is really so, then we might infer that there would be found a characteristic difference of soils in the neighborhoods of different rocks; were it not for the fact, before stated, that all soils have been so transported and mixed, as to preclude the expectation of finding any now remaining unmixed in the region of their formation. When we take this fact into view, I think we may safely conclude that rocks afford but a poor criterion for judging of the character of a soil, and poorer still for deciding upon the treatment best suited to it.

AMENDING SOILS.

93. Most, if not all soils, produce well, when first brought under cultivation. Few, if any, continue to produce well long, unless well managed. These facts

show that more depends upon the farmer *on* a farm, than upon the rocks *under* it.

94. We all know, that where a torrent from the hills flows into a pond, it deposits its gravel at, or a little above, its mouth, while it carries its fine sand into the pond, and its still finer sediment some distance further. If that pond should be drained and cultivated, it is quite possible that the land above the former mouth of the stream might be found too gravelly; that, just below, too sandy; and that, at some distance, too clayey. Various causes, on a larger scale, some of them probably similar to this, have left rather too much coarse matter in some places, too much silica in others, and in some not enough. Energy and persevering labor, scientifically directed, will overcome the difficulties; and nearly all lands will yet be made good. Science has shown that our poorest pine plains have in them the essential elements of grain crops for an indefinitely long time to come; that they only need to be brought into action, and that this *can* be done. We all know that our swamps, now almost useless—better sunk than floating, if that would not make a worse hole than now exists—are sources of endless fertility. We will not blame our fathers, that they did not bring them into cultivation; they could not do everything; but let us do, in this matter, what they (perhaps wisely, in their circumstances) have left undone.

95. Nearly all lands are yet to be made productive. We must take first those that will pay best. Others

will pay by-and-bye. I do not despair of the time, when the man who toils, if he toils intelligently, on a poor farm, will be as well paid as he who works on a good one, after taking into account the rise in the value of the first, and comparing it with the stationary or retrograde value of the other. Thousands of unseemly spots, sand and bog, on which it might have been unwise for our fathers to invest capital fifty years ago, would make an excellent return for capital invested on them this day; and there is every reason to believe, that others will fast come into the same relation to capital and labor—will pay well, ten, fifteen and twenty years hence.

96. In the matter of reclaiming lands, as well as of cultivating those already good, farmers should be guided by experience, by observation, and by common sense. Undoubtedly these are the best teachers. But they are not the only teachers. Science proffers her sympathy and her instructions. Farmers should welcome her aid. Why should they despair of her willingness and her ability to benefit them, when they see what she has done for other interests; manufacturing bales of goods with the labor once required for single pieces; sending merchandise with the speed of steam, and mercantile intelligence with that of lightning?

97. Science has its various branches; and if it be asked, what particular science is most adapted to benefit agriculture, I answer without hesitation, that every science teaches things, which the farmer may turn to

practical use. Zoology has important relations to the rearing of useful animals, and to the destroying of noxious insects. Geology has done much to develop resources, beneficial to all interests; and it deserves especially well of the farmer; it has brought to light fertilizing materials of great value; and it stands ready to teach various lessons, which farmers would do well to hear. But of all the sciences for aiding practical agriculture, *chemistry* is first. The farmer should not only heed what the chemist tells him, but should learn something of this science for himself. It is inwrought with his very employment. The farmer's whole life is spent in performing, or in aiding nature to perform, chemical operations. He should understand how the thing is done. Even when he does right, without knowing why, it would at least be a satisfaction to know THE REASON.

PHYSICAL CONSTITUTION OF SOILS.

98. It has been stated, that the *igneous* rocks (those which had their origin in the action of intense heat) lie below the *aqueous* (those which have been deposited from water). This is true, with the exception of such igneous rocks as have been forced up by volcanic action through the aqueous rocks, and deposited above them. *Granite* is the result of the most ancient volcanic action of which there is now any evidence remaining. Immense quantities of this rock seem to have been forced up in a melted state, forming extensive mountain ranges. Portions of this, as well as of other rocks, have since been broken down, and scat-

tered over the earth's surface, in the form of boulders and pebbles, by what has been termed the drift agency. *Trap-rock*, of which there are two kinds, *basalt* and *greenstone*, seems to have resulted from the volcanic action of a later, but still very ancient period. Mountain ranges of this are also found in various places, as the Holyoke range in Hampshire county. *Lava* is the result of still more recent volcanic action, including that of volcanoes now in existence. The granite, trap-rock, and lava, which appear on or near the surface, are therefore to be considered as having come from deep in the earth. They have been forced up, as lava still is, by volcanic action. Their presence above the aqueous rocks, in such vast quantities, indicates an immense amount of the same materials below them. Next below the aqueous rocks, is supposed to be that vast amount of granite, of which the portions existing on the surface of the earth, as thrown up by ancient volcanoes, are but mere specimens. Next below the granite is supposed to lie the trap-rock, from which the less ancient volcanoes were supplied with the material which they belched forth. Below the trap, geologists believe, is the molten lava, which existing volcanoes are now throwing out. It has been ascertained beyond a doubt, that as we descend into the earth the temperature becomes warmer, at a rate that would bring it to the melting point of rock, at something like forty miles from the surface. It has therefore been inferred that the solid crust of the earth cannot be more than 40 or 50 miles in thickness. This crust, or shell, is supposed to be made up, first, above the lava of trap-rock, then granite; then the aqueous rocks, the primary, the secondary and the tertiary; and then above these the drift and the alluvial deposits. It is not to be supposed, however, that each of these forms an entire, unbroken layer or coating around the whole earth. This is probably true of the igneous formations (the trap and the granite). It is different with the aqueous formations. The primary rocks have been broken in many places, and forced asunder by the ejectment of igneous matter from below. In other places they have been lifted up, by internal heavings of the earth, so high that no secondary rocks have been formed above them. Consequently the secondary formation is more broken than the primary. The tertiary is still more broken, covering but comparatively small portions of the earth. The drift is of very unequal thickness, having been lodged by the agency that distributed it, more in valleys, less on high grounds, and not at all on mountains. The alluvial deposits are of very limited extent, confined mostly to the banks of rivers, which have deposited them; to peat swamps, formed by decaying vegetable matter; and to the slopes and valleys about volcanoes, furnished by volcanic matter from the bowels of the earth. Any deposits, which are the results of causes now in operation, are considered as alluvial. It will be seen from the foregoing statements, that what we call *the soil* (the cultivable portion of the earth's surface, some 10 or 12 inches deep) may lie on either of the aqueous, or stratified rock formations, or even on granite, or trap beds, with nothing but drift intervening. If the soil lie thus above granite, we call that a *granite* region, as New Hampshire; if it lie above the primary, stratified rocks, we call it a *primary* region, as large portions of Massachusetts; if above secondary rocks, a *secondary* region; if above tertiary rocks, *tertiary*; and if above alluvial deposits, *alluvial*, thus naming each district from the underlaying formation.

99. Let us now look at some of the changes which the soil must have undergone. No one can examine it with a powerful microscope without perceiving that it consists principally of rock broken down to various degrees of fineness, from the troublesome boulder to the minutest particle. It bears unmistakable marks of an igneous origin. It must have been once belched from the bosom of the earth in a state of intense ignition. From this state it must have been cooled and solidified. Much of it also bears indubitable marks of having been since broken up, violently agitated by water, and again solidified in the form of stratified rock. From this state it appears to have been again broken up and distributed about the earth in the form of boulders, pebbles, coarse sand, fine sand, and clay. The action of rains and frost has been long at work, rendering it still finer than when first deposited in its present locations. If we could go back to a time when the earth was, in the language of Scripture, "without form and void," or, as it might be translated, "was desolation and emptiness," when as yet no plants had sprung from its surface, we should probably find the materials which now constitute our soils in a comparatively coarse and uncultivable state. In process of time shrubs and trees sprung up. Successive growths lived and perished, d[illegible]ying their

nutriment from deep in the ground, and depositing it on the surface, and thus accumulating and mingling with the surface soil, a rich, vegetable mould. It was in this way, so far as we can judge from present appearances, that the Almighty prepared the soil for his creatures. It was in this state that our fathers received from the Infinite Father the soil of this land. The soil had been formed from comminuted rocks. With it had been mingled a black, carbonaceous mould, extending from a few inches to several feet in depth, and amounting to perhaps from five to fifty per cent. of the whole. The benign, ever-working Power of the universe had thus prepared the soil by such agencies as He chose. Volcanoes, earthquakes, floods, heat and cold, sunshine and shower, successive generations of plants and animals, and we know not what other agencies, had been His servants. He had not made it all a garden. He did not require them to make it so at once. But He had made it capable of becoming a fruitful field with such labor as they could bestow, and ere long, with more labor and skill, of becoming a garden, so fast as the wants of His creatures may require. And it is not too much to say, that the man, who, by skill and industry, is converting the portion allotted him into a garden, is so far doing the will of God. I believe if there is an earthly pleasure more pure, more exalted, and more approved of God than any other, it is that of turning the unseemly waste into a fruitful field, and the fruitful field into a garden, "with every tree that is pleasant to the sight and good for food"—"to dress it and to keep it." What is pleasure if this is not?

CHEMISTRY OF SOILS.

100. One reason why rural employments are not regarded as the most desirable in which man can be engaged, as they seem to have been by our Creator, when He put our first parents into a garden, "to dress it and to keep it," and when he ordained that three fourths of the human race should live by agriculture, is, that labor has been held to be the great and almost the only requisite; and physical labor has been esteemed less honorable than intellectual employment. The truth is, that the employment which combines a manly exercise of both the body and the mind is the most favorable to long life and rational happiness; and such precisely is that of the farmer. The Creator never intended that the farmer's labors should be unreasonably severe, nor that he should thrive by mere hand labor without the exercise of the higher faculties; and He has therefore made his employment such as to require extensive and varied knowledge. One important item of knowledge by which the labor of farming may be diminished and its profits increased, is that of the *chemical composition* of soils.

101. Soils differ essentially in their chemical characters. Some are nearly or quite destitute of several ingredients necessary to fertility. Such are poor soils. Good soils may contain them in very different proportions. My present object is not to state these proportions in any given soil, but rather to take a general view of the causes of fertility as they exist in the soil, and in the rain and air which traverse it.

Let us look at a soil made ready for the hand of industry by those protracted agencies before described, rich in all the elements of fertility, and now cleared and loosened up to a reasonable depth; and let us inquire what are the causes of its productiveness, or what there is in and about that soil, which will make it produce well.

102. As we discuss this question, the learner will do well to turn back to the tables as they are referred to, and refresh his memory with what has been said of the substances there enumerated. Does this soil contain the *elements* mentioned in Table I.? The answer is, Yes, it contains every one of them, and it contains nothing else, or next to nothing; but it does not probably contain a single one of them in their elementary, uncombined state.

103. We will now turn to the *binary* compounds in Table I. Passing by the first as unimportant and not to be found in soils, we come to the second, *sulphuric acid* (SO^3). Our soil will contain 1 per cent. or less of this. It is found by actual analysis to form a small part of all fertile soils. But in warm, sweet soils, none of it is found in its acid or sour state. It is combined with some one or more of the bases (see Table III.), forming a sulphate or sulphates, as with lime, for instance, forming sulphate of lime (gypsum). Next we come to *phosphoric acid* (PO^5, Table I.). We should expect to find from $\frac{1}{5}$ to $\frac{1}{2}$ of 1 per cent. of this, but not in its uncombined state. It exists in all fertile soils, combined with lime and other bases (Ta-

ble III.) as phosphates, and is essential to the production of the cereals, and of all of the sweet, nutritive grasses.

104. *Carbonic acid* (CO^2, Table I.) can hardly be said to be an ingredient of the soil, and yet it exists in nearly all soils in combination with some of the bases (Table III.) as carbonates; and all cultivated soils are always producing it. Whenever vegetable matter burns, its carbon combines with oxygen and forms carbonic acid. The same happens when vegetables decay in such circumstances that air has access to them. Vegetable matter in the soil is thus constantly giving off carbonic acid. A portion of this may be supposed to combine with the bases in the soil, to form carbonates. Much of it goes to feed plants, entering their roots, dissolved in water, or ascending to be taken in through the pores of their leaves. When land lies in fallow through the heat of summer, it cannot be doubted, that much of it escapes into the air and is lost, at least to the owner of that field.

105. *Silicic acid* (SiO^3), or Silica, (quartz, flint, sand) constitutes generally from 60 to 90 per cent. of good soils, and often as much as 95 per cent. of sandy soils. Silica is insoluble in water, but is rendered soluble by alkalies. One effect of ashing land, is to render the silica soluble, so that it can be taken up by the roots of plants. Its office seems to be to afford the stiffening material, for the stalk, straw, husk, and other parts which require to be firm in order to support or protect the seeds.

106. *Nitric acid* (NO^5).—This, like carbonic acid, can hardly be said to be a permanent ingredient of the soil, except as it exists in combination with bases forming nitrates. Rain-water is, however, sometimes impregnated with it, particularly in thunder storms. In highly manured soils, it is formed on the surface, by a direct union of its elements, oxygen and nitrogen. It then combines with bases in the soil, forming nitrates, which may often be seen on the surface, as a kind of white mould. Such an appearance always indicates well for the crops, for the nitrates are easily soluble, and act as stimulants to the growth of plants.

107. *Water* (HO).—The office of this compound, to furnish the moisture required by plants, is too well known to require to be spoken of here. There is another, and most important office of water, which is not so well understood; it is that of dissolving the foods of plants, and carrying them into the plant in a state of limpid solutions. All the foods of plants enter them, either as invisible gases through the leaves, or in a state of perfectly limpid solutions, through the roots. Now water will dissolve in itself and hold in solution 3½ times its bulk of oxygen, once and a half its bulk of nitrogen, once and a half its bulk of hydrogen, once its bulk of carbonic acid, and many times its bulk of ammonia. In this way it conveys these and other nutritious gases as food into the plant. Water also dissolves solid substances, some more and others less, and thus carries them in the form of transparent solutions into the plant, as

food. This office will appear the more important, when we consider that all growing plants perspire largely. They take up large quantities of water from the soil, appropriate to their own growth the nutritive matter dissolved in it, and then throw it off from their leaves, by insensible perspiration. The beneficial effect of irrigating grass lands is probably owing mainly to the fact, that as the water passes over the field, it is constantly absorbing gases from the air and conveying them to the roots of the grass. If the water be impure, as happens with many streams, its impurities operate as fertilizers; and the irrigation may in this way be regarded as a sort of liquid manuring.

108. *Oxides of Iron* (FeO and Fe^2O^3).—The protoxide of iron (FeO) seldom exists in soils, except in those which are low, wet and boggy. This, as before stated, turns to the sesquioxide (Fe^2O^3), under the influence of cultivation. This latter is red, and it is this which gives that color to so many soils. In a rich and productive soil, such as we are now considering, it may be found in proportions varying from 1 or 2 to 6 or 8 per cent.

109. *Oxides of Manganese.*—Of these, there is but one that deserves to be mentioned as a constituent of soils, the black oxide (MnO^2); and this would seldom be found to exceed one-half of one per cent.

110. *Potash* (KO).—One per cent. of potash would be considered an indication of great fertility so far as

this ingredient is concerned. More may exist in some soils, but oftener less. It is generally found combined with carbonic, or some other acid, as a salt of potash.

111. *Soda* (NaO) exists in soils, varying perhaps from one-tenth to one-half of one per cent.

112. *Lime* (CaO).—Some soils contain not less than 8 or 10 per cent. of lime; while a soil may be excellent, and yet not contain more than one per cent. It is generally in combination with sulphuric, phosphoric, and carbonic acids, forming sulphate, phosphate, and carbonate of lime; or with silica, forming silicate of lime.

113. *Magnesia* (MgO).—One per cent. of this would be a large allowance. Soils generally contain much less. More would be injurious rather than otherwise.

114. *Alumina* (Al^2O^3).—This is a fine white powder. It is the basis of clay, which is a silicate of alumina, composed of about 40 per cent. of alumina and 60 of silica. Good soils contain all the way from 2 to 10 per cent. of alumina. Those containing more than 10 are apt to be too adhesive, and those having less than 2 are too porous and open. If a soil is too clayey, it is difficult to cultivate; if not sufficiently clayey, it lacks the power of retaining the food of plants, and allows them to escape by both evaporation and filtration. Many a sandy soil would be more benefited by 10 loads of manure and 10 of clay, than

by 20 of manure; and on the other hand, many clay soils would receive more benefit from 10 loads of manure and 10 of sand, than from 20 of manure. The reason is, that in one case, the clay enables the sandy soil to hold the manure till wanted by the plants; and in the other case, the sand renders the clay soil more light, open, and porous, so that the air can circulate through it.

115. *Chloride of Sodium* (NaCl), or common salt, is found in all good soils, in small quantities, not exceeding 2 or 3 tenths of one per cent. It is oftener exhausted from lands remote from salt water. Lands near the sea are constantly supplied with minute portions of it, in the fogs and rains blown from the sea to the land.

116. *Sulphuret of Iron.*—As before stated, there are three sulphurets of iron, the protosulphuret (FeS), the sesquisulphuret (Fe^2S^3), and the bisulphuret (FeS^2). The first often occurs in boggy and marshy soils. It is not known to be in itself hurtful to vegetation, but when exposed to the air it absorbs oxygen, which coverts the sulphur into sulphuric acid, and this last, combining with the iron, forms sulphate of iron, which is decidedly injurious to vegetation. The injurious effects are counteracted by the use of lime, marl, or ashes. The latter should not be applied till the land is thoroughly drained, as the soluble parts (potash and soda) would otherwise be lost. The bisulphuret is abundant in nature, existing in all rock formations, and probably in nearly all soils. When crystallized, it

takes the color and form of yellow cubes, resembling gold, for which reason it has received the name, as before mentioned, of *fool's gold.*

117. *Sulphuret of Hydrogen* (HS).—This is a gas, having the fetid smell of spoiled eggs. It cannot be regarded as a permanent ingredient of soils, but in richly-manured lands, it is formed in the soil, and may have something to do with the growth of plants.

118. *Light Carburetted Hydrogen* (CH^2).—This is the gas which rises and floats in bubbles on the surface of water, in which vegetable matter is decaying. It is formed also in soils in which there is vegetable matter far below the surface. Vegetable matter decaying in the air, produces carbonic acid; but when decaying with the exclusion of air, it gives off carburetted hydrogen.

119. *Heavy carburetted hydrogen* (C^2H^2) is not known to possess any relations to agriculture. This is the gas used for purposes of lighting.

120. *Ammonia* (NH^3) (see Table I., 20) is a most valuable, though not a permanent ingredient of soils. In conjunction with carbonic acid, it exists in the air, in exceedingly minute quantities, and rain-water and snow are always impregnated with it. More will be said of its relations to the growth of crops hereafter.

121. From what has now been stated, it appears that the mineral part of soils is made up essentially

of the fifteen elements enumerated in Table I., and yet that none of these elements exist in soils in their simple uncombined state; also, that nearly all the compounds in Table I. either constitute a portion of soils, or are in some way so connected with soils, as to act a part in the process of vegetation. These binary compounds, however, very few of them, exist in soils, as binary compounds. They are further combined with each other, forming salts (see Table II.). It must be recollected that the acids combine with the bases (Table III.) and form salts, whose names end in *ate*, the name in each case expressing the compounds of which the salt is formed. If you were to put sulphuric acid and quick-lime into a soil together, they would not remain sulphuric acid and quick-lime. The acid would immediately combine with the lime, and sulphate of lime (gypsum) would be the result. So there are constant changes going on in the soil, and the higher the cultivation, the more rapid and numerous the changes. To control these changes, to arrest such as are unfavorable, and to hasten those which are favorable to the growing of crops, is the great object of scientific agriculture. When this is better understood, the farmer can increase his crops without increasing the expense in an equal proportion, and, consequently, he can increase his profits.

122. Soils consist of two parts—the *organic* and the *inorganic*. By the inorganic we are to understand the mineral part, that which remains after a portion of soil has been heated to redness; by the organic, that which burns away. The organic part is animal and

vegetable matter in process of decay, but not yet wholly decomposed. It always consists of carbon, hydrogen, oxygen, and nitrogen (CHON). In a poor, worn-out soil, there is very little organic matter. In a new and rich soil, such as we have been considering, there is a large amount, sometimes as high as 20 per cent., and very often as high as ten. So much, however, is not necessary, even to the highest fertility. Some of the most productive soils contain not more than two per cent. Organic matter in soils passes through successive changes before it is wholly decomposed into its original elements. At first you will find it in the form of decaying grass, weeds, stubble, leaves, roots, &c. In this state you may sift it out with a coarse sieve. As the process of decay goes on, it takes in oxygen and becomes an acid, as we have seen that sulphur, carbon, and other substances become acids by combining with oxygen. As the process proceeds, it takes more oxygen, and becomes another and different acid. These are called organic acids. Chemists have distinguished no less than five of them—*humic*, *ulmic*, *geic*, *crenic*, and *apocrenic* acids. Others have chosen to call the decaying matter in the soil *geine.* They make this distinction, however, down to that point in the process of decay at which it dissolves in water, they call it *insoluble geine*, and beyond that, *soluble* geine. But as vegetable matter, in process of decay, becomes sour, and then changes its character, becoming a somewhat different substance at each stage in the process, there may be a propriety in calling it an acid, and in giving it a separate name for each stage; and hence the propriety of the names *humic*, *ulmic*, &c.

123. Besides these acids there are also many other vegetable acids. Only two need be mentioned here. One of these is oxalic acid, composed of carbon and oxygen (C^2O^3); the other is acetic acid (vinegar), composed of carbon, oxygen, and hydrogen (CHO). These, together with the five vegetable acids before named, combine with the bases (Table III.), and form compounds named from the acid and the base, in the same manner as the inorganic acids; as acetate of potash, oxalate of lime, &c., thus:

ORGANIC ACIDS.	SALTS.
Oxalic Acid,	Oxalates,
Acetic Acid,	Acetates,
Humic Acid,	Humates,
Ulmic Acid,	Ulmates,
Geic Acid,	Geates,
Crenic Acid,	Crenates,
Apocrenic Acid,	Apocrenates.

CHAPTER III.

VEGETABLE PHYSIOLOGY,

IN ITS RELATIONS TO AGRICULTURE.

GERMINATION OF SEEDS.

124. THE well-matured seed contains in itself the embryo of a new plant, together with food sufficient for the young plant to feed upon, till it shall have had time to thrust its roots into the soil, and its leaves into the air, to draw thence nourishment for itself.

125. This embryo, with its future food closely packed around it, is so snugly encased, generally in a shell or an oily skin, that it will remain dormant, like certain animals in winter, but with undiminished vitality, till the circumstances requisite for calling it into new life are furnished.

126. The embryo, being a perfect plant in miniature, as shown by the microscope, has but to enlarge itself in the directions already commenced, to become a

normal specimen, after the likeness of the parent plant.

127. The germ consists of a *plumule* and *radicle*, the first of which is destined to shoot upward into stem, branches and leaves; the last, to spread itself in the soil into roots. Each leaf is to be an absorbent of vegetable food from the air; and each root, with an open mouth at its extremity, is to run, as fast as possible, after the best food contained in the ground for that particular plant. There is no more doubt that plants exercise choice—select their food—than that cattle prefer sweet grass to sour; though it has been proved, that in some cases, they will take the wrong food, when they cannot get the right, and make themselves sick by it; just as cattle will eat sour grass, when they can get no other, and as men will eat improper food rather than starve. As brutes will suffer more than men, before they will resort to poisonous diet, so there is reason to believe that plants will endure hunger still longer than brutes, before they will take unwholesome food.

128. That they will, in extreme cases, take it, and become sickly in consequence, is now pretty generally conceded; and when therefore you see a stinted, yellow plant, with no worm at its root, nor any visible cause for its misfortune, you may conclude that it is dying a lingering, cruel death, partly by starvation and partly by poison; for it is now pretty well decided that plants, contrary to what was once believed, will absorb poison, before they will quite starve.

REQUISITES OF GERMINATION.

129. While the embryo is sleeping in the parent seed, it has no hold on the earth or air. The circumstances which arouse it to go forth, are *warmth*, *moisture*, and *air*, with absence of light. Its food, till it has grown sufficiently to reach the earth with its roots, and the air with its leaves, must be derived from the seed in which it is shut up. This food consists of starch, gluten, and albumen. Now when you plant a seed, one it may be which has lain dormant ever since the days of the Pharaos, you put it into circumstances requisite for germination—you give it the gentle warmth of the ground, you give it moisture; by covering it lightly, you admit the air, and the air contains oxygen, without which no seed can germinate, nor any plant live, nor any animal breathe; and by covering it to a sufficient depth you partially exclude the light, which is hurtful to the early stages of vegetation.

130. If you had sore eyes you might shrink from the light, though at another time you would rejoice in its genial influences. So a plant, till its first leaves are unfolded, hates the light, but loves it afterwards.

PROCESS OF GERMINATION.

131. When you supply the circumstances requisite to germination, a chemical action commences within the seed, by which heat is evolved. Materials were stored up, ready to act. It is very much as if you

had a stove filled with wood and dry faggots. It may have been so filled a long time. But no heat is evolved. The stove is no warmer than the objects around it. If now you apply a torch, a chemical action takes place in the stove. Oxygen combines with the wood. A transformation of the air and wood into other substances takes place. A real chemical experiment is performed, one that would seem very wonderful, if we had not seen it so often; and much more heat is produced, than was in the torch, which you applied.

132. Just so is it with the seed. There were materials deposited, as in the stove; not to burn, it is true, but to be transformed; and, in the transformation, to evolve heat in the seed, much more than is applied from the soil. As the stove, so the seed, heats itself, when the operation is once started.

Upon this evolution of inward heat, a portion of vinegar is formed in the seed. As cider, by excessive fermentation, turns to vinegar, so a portion of every germinating seed turns into vinegar, or acetic acid. This is believed to attract bases from the surrounding soil, and to form with them *acetates* (123), which are known to be very soluble, and may be regarded as a sort of pap for the embryo plant, while yet it can neither reach after, nor could digest other food.

133. Simultaneously with the formation of vinegar, another substance is formed in the seed, called *diastase.* This diastase has the power to transform starch into sugar. That this is the object there can be no doubt;

for it actually performs this office. In the dry seed there is no sugar. There is starch, a substance familiar to all; there is gluten, the substance which remains in one's teeth after long chewing a kernel of wheat; and there is albumen, a limpid substance, which is recognized in the white of an egg; but there is no sugar.

134. If you taste a corn of wheat in its dry state, you perceive no sweetness; but if you taste it after germination has commenced, you find it sensibly sweet. The same change takes place in cooking flour. The flour, unless it has been damaged, possesses little or no sweetness. But when you wet it, and then bake it, a part of its starch is turned into sugar, and your bread is sweet.

135. As infants delight in sweets, and as the great Designer of all things has caused a peculiar kind of sugar to be dissolved in the food destined for their first nourishment; so the infant plant requires its pap to be sweetened, and the wise Designer has made provision for the exigency. True, he has not deposited sugar in the seed; for sugar, being soluble, would be dissolved, and washed out by the winter rains; but instead of sugar, which is soluble, and consequently not permanent, he has deposited starch, which is insoluble and somewhat permanent; and has at the same time made provision for its transformation into sugar, through the agency of diastase, at the very time when wanted by the young plant.

136. It is manifest that the production of heat in the germinating seed; the formation of vinegar and diastase; and the transformation, by the latter, of starch into sugar, are all provisions of that Being who is wonderful in counsel, for the express purpose of furnishing suitable food to infant plants, when they could not obtain it otherwise; and, per consequence, of providing abundant food for man and beast.

137. There is another fact worthy of reflection. It has been proved by the most accurate experiments, that seeds, during their germination, and up to the time of their first putting forth leaves, absorb oxygen and emit carbonic acid, the reverse of what takes place subsequently. Now why is this? Probably that the embryo plant may be surrounded with carbon, dissolved in the water of the soil, and may thus obtain through its first roots, that kind of food, carbon, which it is destined subsequently to receive from the air through its leaves. This seems very much like a provision for it, on its way up into the air, not unlike what would happen, if a mother, whose son was starting for a long and solitary walk, should slip into his pocket some food for the way. Every one can make his own reflections. To me the fact seems worthy of notice.

GROWTH OF PLANTS.

138. You can hardly have failed to reflect, that much care has been bestowed by the Divine Architect to give the plant a good start into being. The husband-

man, who will exercise a like care, that his plants commence well, will be so far a co-worker with God. Plants should not be so puny for a month after they are up, that, if a worm or a bug take a mouthful from them, he will take the whole. By a prudent forecast, in preparing the ground and the seeds properly, and in selecting a suitable time for planting, we should endeavor to give them a good start. We should use forethought, and take special care for their infancy. More than is generally considered depends upon giving our plants a good setting out on their summer's career. If this is not the whole of the battle, it is certainly an important part of it.

139. I do not mean to say that by due care of their infancy you can make them so powerful that they will compete successfully with poke and pig-weed for the food of the soil; or be able to resist the encroachments of horned-cattle and swine; but I will say, that by starting them vigorously, you can make them put forth brawny arms, long roots, and broad leaves, by which to draw for their productiveness from sources which cost you nothing—from the air and from the subsoil.

140. It should be remembered that a portion of that which makes plants grow, is at our own disposal, as our soils and our manures; while another and about an equal portion is in common stock, blown about by the winds of heaven. Now if we work rightly that which is at our own disposal; if we make our soils deep, mellow and friable; if we put in the manures, instead of

letting them steam away, or wash off from about our dwellings, polluting the air we breathe, and perhaps sooner or later the water we drink; if we let no giant weeds filch the food in our fields; we shall draw more largely from the common stock; for we make our plants more vigorous and far-reaching and successful in their efforts to draw from the great store-house of vegetable food above and around us.

141. This is one of the ways in which Divine Providence rewards the diligent and punishes the slothful. The thorough farmer, by high cultivation, gets a great deal more out of the common stock, than the mere ordinary farmer. Not all the corn comes from the soil; not all from the soil and manure together; half of it comes from sources, which cost nothing, as free as the breezes of heaven; one acre well tilled draws more from the common stock of corn-making materials, than two acres half tilled; and the net profit on one acre highly cultivated is more than on five, that are barely run over.

142. We have all heard of the *dish being right side up*. When the farmer's field is mellowed to a depth of 8, 10, or 12 inches; when the crops are running their roots deep and their tops high; when every leaf and every inch of surface soil are sucking in the rains, and dews, and nutritious gases; then is his dish right side up; and he will catch enough, not only to pay him for his labor, but to give him a handsome profit.

GROWING PLANTS PURIFY THE AIR.

143. When a plant has put forth its first leaves, and is no longer dependent on the seed for its support, it reverses the process before described—absorbs carbonic acid and emits oxygen, during the day and so long as light continues, but still absorbs oxygen and emits carbonic acid in the night. The carbonic acid is decomposed in the plant, and its carbon wrought into the solid texture of the plant, while its oxygen is given off. Other floating gases are taken into the soil and conveyed to the plant through its roots. Thus growing plants purify the air of those gases which render it unhealthy for respiration; while the respiration of men and beasts enriches it with those gases which promote vegetation; so that plants and animals are mutually beneficial, each rendering the air health-giving to the other. None breathe so invigorating an atmosphere, as the farmer among his growing crops.

SOURCES OF CARBON AND OTHER FOOD TO PLANTS.

144. During the growth of the plant it takes its carbon mainly from the air. A little is believed by physiologists to pass in through the roots, dissolved in water. Its oxygen and hydrogen are undoubtedly furnished mostly in the form of water, and in that form taken in both by the roots and leaves.

145. Nitrogen is furnished to plants principally in the form of nitric acid and ammonia, both of which exist in the air and in rain-water.

146. So fai as the four organic elements are concerned, the plant obtains them from the air mainly, either directly by the leaves, or through the surface soil by the roots.

147. It would not be far from the truth to say that the plant feeds itself about equally from the earth and the air during its growth. Its inorganic matter, that which remains as ash, when the plant is burnt, is obtained wholly from the ground, but is only a small part of the whole, not more than from one to ten per cent. It is probable that a poor, stinted crop is derived from the soil and air in about the same proportions as a luxuriant one. But the whole of such a crop is a small affair. A part of it is still smaller; and I wish here to repeat and impress the thought, that the better we do by our plants, in their ground relations, the more they draw for us from the common stock of vegetable food, which floats unseen in the air.

FLOWERING AND SEED-BEARING OF PLANTS.

148. One thing should be noticed with regard to the flowering of plants. The flower-leaves, unlike those of the other parts of the plant, absorb oxygen by day as well as by night. The object of this arrangement probably is to give them their beautiful colors. The oxidizing of various substances changes their hue. For instance, if a flower-leaf have in it a trace of the protoxide of iron, the inhaling of oxygen will give it a brilliant red. Other substances are turned by the same cause into blue, yellow, violet, &c.

149. As plants approach their seed-time, their principal effort seems to be concentrated upon the one object of maturing seed. With many plants, especially with the cereals, I suppose it to be a well-known fact, that this function is sometimes performed better than their previous growth would lead one to expect, at others not as well—that the growth is not to be taken in all cases, as a measure of the fruitfulness. If the fruitfulness exceeds the growth, we may safely conclude that the ground is better supplied with the requisites for maturing seed, than with those for promoting growth. If the growth exceeds the fruitfulness, we may suppose the contrary to be true; provided in both cases no other cause appears, by which the disparity can be accounted for. The causes for the failure of crops in their last stage, are undoubtedly various. Sometimes it is attributable to the season; the early part of summer being favorable to luxuriant growth; the latter, unpropitious to the maturing of the seed. Oftener, I believe, it is owing to some mismanagement. With regard to the corn crop, I have always thought that the putting of a little stimulating manure in the hill, without thoroughly pulverizing and enriching the whole field, was precisely adapted to produce a large growth of stalks with little corn. I should anticipate that the effect of such manuring would cease at the wrong time, not solely from the exhaustion of the manure, but because it was confined to one place, instead of being diffused through the soil. Corn roots do not curl down under the hill; they spread over the field as widely and as deeply as the ground has been prepared to receive them. Why

should the manure be in one place—immediately under the hill—unless you mean to discourage the roots from taking a broad range in search of food?

LATE HOEING INJURIOUS.

150. I need not say here that another bad practice is that of letting the weeds live and compete with the corn for the strength of the soil, for I suppose no such practice obtains among us. But there is a practice, not much better, which prevails in many places, that of killing the weeds at so late a period as nearly to kill the corn too. I have seen men hoeing corn at so late a day, that if the corn had been mine, I would have thanked them heartily to let it alone.

151. If you cut off the roots of a tree it will send out two new roots for every one that is cut off, and the tree may not be injured. Some think it will become more vigorous. But if you cut the roots of corn, after it has silked out, and thus force it into the business of forming new roots, at the very time when it should be maturing its seeds, you commit a fatal mistake. You might just about as well bleed your horse half to death, and work him hard in order to fatten him, especially if you would keep him rather short the while, as corn is of course kept short, while it has few unmutilated roots to convey it food.

STRUCTURE AND CIRCULATION OF PLANTS.

152. Of the structure and circulation of plants I

have space to say but little, as the more important matter of their decay and return to the soil is yet untouched.

153. With regard to the structure of plants, I will refer you to our common trees, not exactly as a sample for others, but as affording some data from which you can reason, and observe for yourselves both resemblances and differences.

154. The stem of a tree consists of woody fibre, formed around the pith, an inner bark around that, and an outer bark around the whole.

155. The pith is a spongy, soft substance, commencing far down in the roots, coming together at the base of the stem, then continuing upward, dividing and subdividing itself in the branches and twigs, till it reaches their extremities. Some mysterious connection seems to be kept up between the pith and the inner bark, by means of a set of pores, running from the pith outward every way, like the spokes of a wheel.

156. The roots may be regarded as the extension of the stem downwards, and the branches as its extension upwards. The wood is not as compact as many may suppose. Its more solid parts even consist of an immense number of tubes running side by side from the lower to the upper extremities of the tree, varying in size in different parts, and each one lined with a substance different from itself, like the tinning

of an iron kettle. It is through these that the sap passes upwards.

157. If we commence at the extremities of the roots and examine, we shall find that the extremities, called *spongioles*, consist only of a bark and a porous substance enclosed. This porous substance extends out to the end of the bark, and is adapted to the absorption of water and watery solutions. Nothing enters a tree or other plant that is not perfectly dissolved, as limpid and transparent as the solution of an ounce of salt or sugar in a gallon of pure water.

158. If we trace the rootlets from the spongioles upward, we shall find them gradually increasing in size and hardness, and coming together, till instead of millions, there will be only a few; becoming more and more like the wood of the trunk; and before reaching the stem, invested like the tree itself with a double bark, and having like that a pith in the centre.

159. As we ascend we shall find the branches and twigs becoming more porous as they recede from the stem. Were we to burn the small branches and leaves, we should find them to contain three or four times as much ash as the solid wood, and of the best quality.

160. The leaf-stems are a continuation of the twig. They are bundles of tubes enclosed in bark; and these tubes connect with those of the wood below.

Through these, the sap, which may be regarded as the blood of the tree, flows upward into the leaves. The leaves may be compared to the lungs of animals. The office of the former is to bring the sap and the air into contact, as that of the latter is to bring the air into contact with the blood. As the blood is strikingly modified and changed in the lungs, so is the sap in the leaves. The lungs are a net-work of blood and air vessels surrounded by a membranous tissue. So also the leaves are a net-work of woody fibre, continued from the leaf-stem, and covered above and below with a spongy membrane. The upper side of the leaf emits gases and vapor into the air; the under side gathers in from the air for the nourishment of the plant.

161. When the sap has circulated through the leaf, it commences a retrograde course towards the earth. It is not always a downward course. That depends upon the position of the limbs. Its return to the earth is by the inner bark; and its depositions by the way form the annual layer of wood.

DECAY AND PRODUCTS OF PLANTS.

162. In the present order of things, whatever lives, must die. Men, brutes, and plants, live on their predecessors. The floating matter of the universe is undergoing a succession of life and death. Probably all the dead matter around us, all that we can see, has been alive some time, much of it a thousand times. The succession of living beings, vegetable and animal,

is kept up by using the same matter over and over again.

163. The plant, in its growth, devours other plants, and even animals; for it finds no richer food than dead animal matter; but in its turn, it is destined to be devoured either by animals or plants, and pretty surely by both. A particle of dead matter now in the soil may be clover next summer, beef next winter, and clover again in six months. There is a restless activity in the matter which composes the surface of the earth and its surroundings.

164. When plants have passed their maturity, they yield, among their earliest products of decay, called proximate constituents, wood, starch, gum, sugar, gluten, caseine, and albumen. Out of these grow the secondary products, alcohol, vinegar, and too many others to be named. The secondary products of decay are counted by thousands and hundreds of thousands, if not by millions. Notice cannot be taken of them here. But those primary products which I have named are of great importance, as forming, directly or indirectly, almost the entire food for the human race, and for all the animals that live.

165. *Starch* is of course pretty well known. It is, however, known by different names, as it is derived from different plants, as potato starch, wheat starch, &c. Sometimes it takes the name of the country whence it either is, or professes to be, imported, as Poland starch. That which is obtained from the pith

of the palm tree is called in commerce *Sago;* that from the roots of the Maranta arundinacea of the West Indies is known as *Arrow-root;* and that from the root of the manioc tree is the well-known *tapioca* of the shops. All these—starch, sago, arrow-root, and tapioca—are substantially the same thing. They are all washed with cold water from the substances in which they are respectively found.

166. Starch, gum, and sugar contain no nitrogen They are all characterized by the letters CHO, signifying carbon, hydrogen, and oxygen.

167. On the other hand, gluten, caseine, and albumen are *nitrogenous* substances. All of them contain nitrogen, and all contain sulphur and a very little phosphorus. Their principal ingredients being carbon, hydrogen, oxygen, and nitrogen, they are characterized by the letters CHON, expressive of their composition.

168. One or more of these nitrogenous substances exist in all plants.

169. *Gluten* is a very important constituent of wheat. It is insoluble in water. Hence, if you chew a kernel of wheat, the gluten will remain in the mouth after the rest will have disappeared; or, if you wash wheat flour over a cloth, the gluten will remain on the cloth, a tough, stringy, grayish substance, while the starch and the albumen will pass through.

The gluten is the most nourishing part of wheat; and that wheat is best which contains most of it.

170. *Caseine*, which strongly resembles curd, is found abundantly in peas and beans. It is soluble in water, and will coagulate, like the curd in milk, if an acid, as vinegar or rennet, be added.

171. *Albumen* abounds in oily seeds, as poppy seed, flax-seed, &c. It is soluble in water, but coagulates, like the white of eggs, if boiled.

172. You can easily separate the constituents of flour, and examine them in the following manner:

173. Wash two or three ounces of fine flour on a piece of linen or cotton cloth of medium thickness, with as many pints of water. Pour on the water, a little at a time, and stir the flour gently on the cloth, letting the water fall into a pan below. What remains on the cloth is gluten. After the water has stood in the pan long enough to become perfectly clear, pour it into a kettle so gently as not to disturb the sediment. What remains in the pan is starch. Then heat the kettle till the water boils, and the albumen will be seen in a coagulated state, having some resemblance to the white of an egg after being partially boiled.

174. I have already stated that gluten, albumen, and caseine are nitrogenous substances, and that they

contain a little sulphur and a trace of phosphorus (CHONSP).

175. Sugar, gum, and starch, on the other hand, contain no nitrogen, and are therefore less nutritious as articles of food. The elements are the same in each, and in the same proportion as represented below:

Starch, - - - - -	$C^{12}H^{10}O^{10}$
Gum, - - - - -	$C^{12}H^{10}O^{10}$
Sugar, - - - - -	$C^{12}H^{10}O^{10}$

It will be seen that the oxygen and hydrogen in these three substances exist in the same proportion as in water. The same is true of woody fibre and of many other substances. They consist of carbon and the elements of water. I ought perhaps to state that the sugar before characterized is cane sugar. There are other kinds of sugar. Grape sugar, for instance, is differently constituted; the sugar of milk is still different; and that of the ash tree (the manna of commerce) is different from either.

TRANSFORMATIONS.

176. The fact that starch, sugar, and gum are the same in chemical constitution, and that they are transformable one into the other, is one of the most remarkable discoveries of the last half century. That they are identical in constitution, and that they are transformable, is quite certain. The transformations are actually going on in the operations of nature, as

in germinating seeds, in the sap of the maple tree, and in ripening fruits. And, what is more, these transformations can be imitated by the chemist. You may take an ounce of starch and turn it all into gum; you may then turn this gum into sugar. Nor need you stop here; you may dissolve this sugar in water, and then, if you expose it to air and warmth, and add to it a single particle of yeast, you will transform it into alcohol. And you need not stop here; for, if you let the fermentation proceed one step farther, you transform the alcohol into vinegar. You cannot change starch into vinegar directly, but you can do it by the route I have described. You may even go back one step farther, and commence with woody fibre. You may change woody fibre into starch. Your routine of transformations then would be, woody fibre, starch, gum, sugar, alcohol, vinegar. It should be remarked, however, that the constitution of the two last becomes changed. Alcohol and vinegar are of the same elements, but not in the same proportions as the others.

177. Such are some of the proximate constituents of plants, and a few of the numerous and wonderful transformations to which they are subject.

178. Sooner or later all these substances, which plants have so curiously elaborated out of dead matter, are destroyed. Their organic elements return to the air; their inorganic, to the soil; both to the place whence they came; both, as undistinguished atoms, to be used in building up new plants and new animals, which, in their turn, are to perish and become

food for others still; and so on in successive rounds, just so long as the Almighty Worker of the universe shall decide to uphold the current order of things. The plants and animals, including men, that have been in ages past, live in those that now are; and those that now are, will live in those yet to be. *Death is the parent of life.*

CHAPTER IV

ANIMALS AND THEIR PRODUCTS.

CONNECTION BETWEEN SOILS, PLANTS, AND ANIMALS.

179. In a former part of this work were described about 20 substances, all of which, either as permanent ingredients of the soil, or less permanently connected with it, contribute to the growth of plants. It may now be stated that the permanent ingredients of soils are fewer in number. Soils are substantially made up of organic matter, *potash*, *soda*, *lime*, *magnesia*, *oxides of iron*, *oxide of manganese*, *sulphuric acid*, *phosphoric acid*, *carbonic acid*, *chlorine*, *silica and alumina*.

180. These twelve constitute soils. If we omit the last, the remaining eleven constitute plants; and if we strike off the last two, the remaining ten constitute animals. Alumina stops in the soil; silica, except in exceedingly minute quantities, stops with the plant; the other ten pass from the soil into the plant; then from the plant into the animal; and finally back into the soil. From this it will be seen that when we ex-

pend crops on the farm, we return to the soil all we took from it, and as much more as the growing plants draw from the air, which is nearly all their organic matter. In this way a farm should be constantly gaining in fertility; for on the supposition that we sell nothing from the farm, we keep all the inorganic parts of the soil at home, and by means of growing plants we are all the while gathering inorganic matter from the air and incorporating it with the soil; so that the soil, treated thus, would remain equally rich in the inorganic (mineral) parts, and be growing every year richer in the organic parts. It will be seen also, that if we sell off crops, or anything that is made from crops, as beef, pork, butter, cheese, the soil must be from that time becoming poorer in the inorganic ingredients, unless we procure fertilizers from off the farm and substitute them for those which we send away; for when we sell any product of the farm, we sell a part of the soil; not enough in a single pound of butter to diminish sensibly the quantity left, but enough in a century, in all the butter that may be sold from cows fed on a single pasture, to leave that pasture entirely destitute of certain ingredients, without which good butter cannot be made. So if the hay from a mowing was to be sold off for many years and nothing returned, certain ingredients of the soil would become so exhausted, that little or no more hay could be grown on that soil; or if the corn, wheat, or rye were to be sold from a soil, the result would be the same. If a soil were eminently good, it would resist bad treatment a long time, but sooner or later it would be exhausted. The farmer who should have sold all his

crops for a long time, and put nothing back, would find that he had sold his farm also—sold it piece-meal.

PREVENTION.

181. To prevent so sad a result, two modes are resorted to. One is to procure foreign fertilizers enough to make a full substitute for what is carried off from the farm. It may be wise to adopt this course near large cities, where produce is always high, and where various fertilizers can be bought cheaply and conveyed to the farm with a small expense—brought home, perhaps, by the same team which draws the produce to market. But with the great mass of farmers, the other mode commends itself as the only one applicable in their case. It is, *to expend their produce mainly on the farm, to preserve every particle of manure, and to compost it with peat, road-scrapings, &c.*, so as greatly to increase the quantity, and *to keep the quality up by adding plaster, salt, lime and ashes.*

182. In this way, a farm can be made increasingly fertile; and if the best animals be selected, and the best modes of feeding be adopted, the annual income of the farm can be made nearly as great as by selling the hay, grain and roots. If, on the other hand, the animals are of inferior qualities, and the modes of feeding are wasteful and ill suited to the nature of the animal, not more than half of the estimated value of crops are returned in the growth, products and labor of the animals that consume them. These considera-

tions are important to the practical farmer. His object in feeding is to keep the land productive, and at the same time to get the value of the produce in the increased worth of whatever consumes it.

KINDS OF ANIMALS TO BE KEPT.

183. The animals usually kept among us are horses, horned cattle, sheep, swine and poultry. It is worthy of inquiry, whether mules ought not in some cases to be added to this list. The arguments in favor of it are, that the mule is very hardy, little liable to disease, capable of thriving on coarse food, requiring less food than the horse, long-lived, and able to perform great labor. For these reasons, it would seem that for some purposes the labor of mules might be advantageously substituted for that of the horse. But for general purposes, including the transportation of persons, the horse must remain in favor; and it may be laid down that horses, horned cattle, sheep and swine, are the animals to consume mainly the produce of American farmers.

184. Animals may be distributed, with regard to the return they make to the owner for the food and care given them, into three classes: those which return labor only, those which return both labor and the products of their bodies, and those which return the products of their bodies only. To the first class belong the horse and the mule. There are some men, perhaps, who have the skill and address to make these animals do work enough to pay for their care and

feed, and at the same time increase in value. But as a general rule, if the farmer is to be paid the worth of the hay and grain they consume, he must take it out in work; and hence, with the exception of those who make it a business to prepare these animals for the market, and of some others, who can afford to keep fine horses for the pleasure of driving them, it is unquestionably a good rule, to keep no more of these than can be pretty constantly employed.

185. Working oxen belong to the second class The return they make the farmer is labor and growth. Like the horse, they can be made profitable, as workers; and unlike the horse, they may be profitable, as idlers. Perhaps few animals pay better for their keeping, than oxen, worked reasonably till August, and then turned into a good pasture to be prepared for the stall. They should seldom be put to the utmost of their strength. If worked with judgment, and kindly cared for, they will do more for their owner in the long run, and will be steadily increasing in value, till 7 or 8 years of age, when they should be prepared for the market, and their place supplied by those that are younger.

186. To the third class belong cows, young stock, sheep, and swine. The return expected from these is the body of the animal, when grown and fattened, or some product of the animal; as butter, cheese and wool. Every farmer knows that some animals will eat much and grow little, while others will consume less and grow more; he knows that there are some cows, which

will give little value in milk, while others will give much; that some sheep give almost valueless fleeces, others valuable ones; and that some breeds of swine are all-consuming, but ever lean. It makes a wide difference, whether food be thrown out at random, or be given with regularity and discretion; and whether animals be made comfortable by adapting circumstances to their several natures, or be left to continual suffering. The farmer knows, or may know, that if he selects the consumers of his produce wisely and keeps them properly, they will give him a return of some 10 dollars a ton for his hay, 6 per cent. or more for the value of his pastures; and a fair remuneration for his corn and roots, together with a moderate compensation for his care for them; while if he neglects these conditions, he gets but half the value of his hay, grain, roots and pasturing, and nothing for his trouble, unless it be the pleasure of railing and complaining that his is an unprofitable business.

GENERAL TREATMENT OF ANIMALS.

187. The conditions of farming are absolute and unalterable. We have seen that the farmer cannot sell the products of his fields abroad. He must dispose of them mainly at home. The inmates of his barn, fold, and sty, must be his pay-masters. They are "good pay" if he manages wisely. If he does not, he has no right to complain, and thus sink his profession in the estimation of his sons and his neighbors. As the farmer's animals are his customers and his pay-masters, he should use them well *for his own sake*. There is a

higher motive for using them well. God has made them sensitive beings, capable of gratitude and of resentment, of great enjoyment and of intense suffering. The gift of them, as such, to man, implies that they are to be treated *kindly*. He who treats them otherwise *offends his Maker*. To inflict needless pain upon a brute, or to make him less a creature of enjoyment than he is capable of being, consistently with our own interest, is, to say the least, to be brutish. High moral obligation concurs with our own interest, in requiring at our hand a considerate, judicious, kind treatment of domestic animals.

188. We should be observant of the habits of those animals, and attentive to their wants. If you see a cow gnawing a bone, you may depend upon it, a sharp necessity impels her to it; you have drawn phosphate of lime from her system in milk till she feels an indescribable longing for that substance. She at least cannot describe it, except by the action of trying hour after hour to masticate a bone. It is probably a feeling, or rather a want, not unlike that of the drunkard, when the state of his system is such that he would almost barter soul and body for a taste of spirit. There is reason to believe that milch cows are often great sufferers, when kept in old worn-out pastures, for the want of phosphate of lime. It should be given to them in the form of bone-dust. If a cow or an ox is seen licking the ground, it indicates a want of something (in most cases salt), which should be given in a more eatable form. So there are a thousand indications, which the attentive farmer will observe; and

from them he will learn the wants of his animals, and learn to supply them.

189. It is almost as important that animals should be kept warm in winter, and have cool shades to resort to in summer, as that they should have plenty of food. The food that will keep one of them growing, when comfortable, will at best only keep him stationary, if suffering with cold. So it is with regard to other matters of comfort. A swine, for instance, finds it very comfortable to wallow in filth. He should be provided for, accordingly. But he finds it very uncomfortable to be confined to such lodgings; and if so confined, he will not pay for his keeping. No animal loves better to retire to a snug, dry nest. This should always be furnished for him. A swine that is kept shivering in his own filth, will eat out during one of our long cold winters twice what his body will be worth in the spring, when if he had been kept warm and dry, he would have paid for his keeping, and the owner might have sold his pork at a fair profit. Perhaps no maxim is more manifestly true, than that duty and interest both concur in requiring a kind attention to the wants of domestic animals. The individual who is slow to perceive, and slower to minister to these wants, will find the inmates of the farm-yard *poor paymasters.*

THE FEEDING OF ANIMALS.

190. Common salt is essential to all animals. As it forms a constituent of soils, and enters from the soil

into plants, it is thus furnished, in minute portions, to domestic animals. But all experience has shown that they require more salt than is thus furnished. This is indicated by the intense hankering they manifest, when long deprived of it. How much they need, beyond that naturally contained in their food, no one has been able to decide. The animals themselves know best, and it should therefore be left to their own choice. If not deprived of it too long, they will always eat just as much as is for the interest of the owner that they should; and no precaution is necessary, except that when a neglected animal is brought home, he should not be admitted to a full supply at once.

191. Some have supposed that cattle while being fattened on hay and roots, especially on potatoes, should be allowed to drink but little. Whether this opinion is well founded, I very much doubt. In all other cases, cattle and sheep should have water plentifully, and where they can get it without fear or danger. If possible, it should be pure fresh water; and should be in the barn-yard.

192. The foregoing remarks apply equally to all kinds of stock. With regard to solid food, a difference should be made, accordingly as animals are kept for one or another purpose. As a general rule, the finest and earliest cut hay should be given to milch cows—those in milk at the time;—and they should have as much as they will eat. Good, substantial hay, not late-cut, nor in the least smoky, should be given to working cattle and horses. Dry cows may be turned

off with poorer hay in part; and young stock should have a variety of food. Very few animals, if kept through the winter on good hay, worth perhaps 10 dollars a ton, will pay for the hay which they consume, in their increased value in the spring.

193. As stock cattle generally rise more or less during the winter, their increased value in the spring must depend partly on their growth and partly on the rise of this kind of property. Thus, if a steer weigh eight cwt. in the fall, and is worth four dollars the cwt. live weight, his full value would be thirty-two dollars. If he weigh nine cwt. the next spring, and be worth five dollars the cwt., his spring value would be forty-five dollars, giving an increase of thirteen dollars. Now, this is a high estimate of increase in value, and yet at this rate the owner would not have realized the value of hay which that steer would have consumed by considerable, provided he had been kept on good hay and nothing else. It may be put down as a settled point, that stock (including oxen not at work, cows not giving milk, steers, heifers, and calves), if kept on hay only, do not, as a general thing, pay for their wintering. The average increase in the value of the stock is but about half the estimated value of the hay. This has hitherto been a sad *leak* in farming. The question is, must it continue?

194. A farmer puts up fifty tons of hay; in the fall it would have sold for $500; his cattle eat it, and are worth $250 more in the spring than the fall before. The manure pays him well for the labor of feeding,

and not much more. There is then a loss of $250. This is a terrible drawback on the profits of farming. We suppose that horses and working oxen pay for their wintering in work. Milch cows pay for their keeping in milk. But stock cattle, fed on hay only, pay in their increased value in the spring for only half the hay they consume. If fed in the very best manner, and kept as warm and comfortable as could be desired, they require at least 2 per cent. of their live weight of good hay daily, in order to keep them in a thriving condition; and the cost of this, at $10 a ton, is more than the average increase in value by nearly one-half. This loss, which has long been felt and bitterly complained of, must be avoided by resorting to mixed food. One pound of Indian meal is of about equal value for feeding with 4 lb. of good hay, 6 lb. of second quality hay, 8 lb. of oat straw cut, 10 lb. of carrots, and 16 lb. of turnips.

195. Here are six kinds of food for cattle, to which may be added corn-stalks, making seven. No one of the seven can be fed alone to growing cattle with paying results. But it does not follow that a just interspersion of the whole may not pay well. A good housewife often sets before her family costly dishes, but she takes care to vary them, and to intersperse such as are less expensive, but so well "got up," and with all so timely, that they may be acceptable. So it is with the wise feeder. Suppose he has twenty-five head of stock cattle, averaging 9 or 10 cwt. live weight each. If he feed them on the best of hay three times each day, he will find that it requires 5 cwt., per day,

to keep them in a thriving condition. This, in 150 days, would amount to 37½ tons, worth, at least, $375, and he will soon find, if he did not know it before, that his cattle will not gain $375 in value when fed in this way—probably not more than half that amount, except in those years when stock happens to be very much higher in the spring than the fall before. Three years out of four he will make a heavy loss on his hay. The problem is, how to avoid this loss. If he can keep his cattle more cheaply, and have them grow equally well, he will gain in one direction; if he can keep them equally cheaply, and have them grow better, he will gain in another direction: if he can gain both these ends—keep them more cheaply, and have them grow faster—then he will gain in both directions. These are the points at which the feeder of stock should aim. He must take the best *care* of his cattle, and give them a *variety* of food. Let them have pure water; let it be where they can get it without much trouble; they should drink little and often, rather than drink enough at once for twenty-four hours. Let them have salt always within their reach. Let them have warm stalls, and a sunny yard, well littered, and thoroughly protected from all cold winds. If, with these conditions, he will give them a little good hay, daily, they will be so hardy, and so contented, that they may be turned off with cheaper food for the principal part of their living, such as second-rate hay, cut straw, with a little corn-meal thrown upon it, roots, corn-stalks, salt hay, swamp grass, almost anything that can by possibility be eaten. If they are to be turned off with much coarse fodder and little fine hay, they should, by all means, have roots, as carrots and turnips, or, if these cannot be furnished, Indian meal. The succulent qualities of roots afford a sort of set-off for the dryness of husks, straw, and poor hay; and corn-meal contains some 10 per cent. of oil, which helps down coarse fodder, about as well as butter does dry bread.

196. No certain rules can be given for all cases. The feeder of stock knows what fodder he has to dispose of, and what sort of cattle he has to consume it. *He must cater* to his cattle as best he can out of his resources. He must carefully note the effect of his feeding from time to time. If the profits of agriculture are to be increased, or, in other words, if the farmer is not only to *raise good crops*, but to *get paid for them*, the business of feeding must receive far greater attention than it yet has.

197. It has been stated in another part of this work, that among the proximate constituents of plants used as food, are woody fibre, starch, gum, sugar, and oil, called non-nitrogenous substances. These are composed of carbon, hydrogen, and oxygen (CHO). There are also three nitrogenous substances—gluten, albumen, and caseine. These, in addition to carbon, hydrogen, and oxygen, contain also nitrogen (CHON). The first mentioned substance, woody fibre, has little to do with nutrition. It passes the animal mostly undigested. The office of the next three, starch, gum, and sugar, is to support respiration. The next substance, oil, goes to form the fat of animals, and the

last three, the nitrogenous substances, furnish material for muscles (lean meat), tendons (cords), and cartilage (gristle). Thus, these organic substances support respiration, and furnish material for all the soft parts of the body. The bones are formed from phosphate of lime, an inorganic substance contained in all the nutritious grasses and in the grains.

198. It has been stated that starch, gum, and sugar, as parts of animal food, go to support respiration. This requires to be explained. When wood is consumed, the oxygen of the air combines with the carbon of the wood, forming carbonic acid (CO^2), and the oxygen and hydrogen of the wood combine with each other, forming water (HO), so that carbonic acid and watery vapor pass off, while the inorganic parts of the wood fall to the hearth in the form of ash. If we were to burn a handful of corn, the same would take place. The organic part would pass off, as carbonic acid and water; and the inorganic part would fall down, as ashes. But if, instead of being burnt, the corn were to be eaten by an ox or other animal, let us see what would become of it. It contains, among other things, starch, gluten, oil, and phosphate of lime. The digestible parts would be first converted into chyle and then into blood. The blood would be immediately forced through the lungs. In the lungs air would come in contact with it. The oxygen of the air combines with the carbon of the starch, and forms carbonic acid; while the other elements of the starch combine with each other, forming water; and the animal exhales carbonic acid and watery vapor, the

same substances that would pass up the chimney, if the corn were burnt on a hearth. And as heat would be produced by the latter operation, and diffused through the room, so animal heat is produced and diffused through the system by the former. Starch undergoes the same changes in the lungs of animals as when thrown upon burning coals—in both cases it is converted into carbonic acid (CO^2), and watery vapor (HO); in both cases these are sent afloat in the surrounding air; and in both cases the same heating effects are exhibited.

199. The lungs may not inaptly be compared to a stove constantly burning; for as the stove converts fuel into carbonic acid and water, and thereby generates heat, so the lungs convert starch, sugar, or gum, which may be regarded as the fuel of the animal system, into the same substances, and thereby generate heat and diffuse it through the system.

200. This explains why animals should be kept warm in order to grow; for if they are exposed to severe cold, it takes a large portion of their food to keep the lungs in sufficiently active operation to prevent their freezing. Little is left to supply the natural waste of the body, and perhaps none at all to furnish material for new growth. It explains also why working animals require more food than others. In consequence of exercise they breathe more, and a larger proportion of their food is exhaled from the lungs. If you were to give a young, thrifty ox, weighing 1,000 lbs., live weight, 25 lbs. of good hay daily, and to keep

him warm and quiet, you might expect him to grow rapidly, because the 25 lbs. of hay would furnish food for his lungs, and leave more than enough to supply the natural waste of the body, and this surplus would go to form new growth. If you were to give the same feed to a similar ox, bu. were to work him 12 hours a day, you could hardly expect him to grow much; for, being in such constant exercise, he would breathe a great deal, and the food might no more than supply fuel for his lungs, and leave enough to make up for the natural waste of the body. Enough for this latter purpose might not be left, and then he would fall away.

201. If you were to give the same food to a third ox similar to the other two, but were to turn him out into a bleak lot covered with snow, with no shelter whatever, you might calculate for a certainty that he would lose weight rapidly. The food, though enough to make him grow, if he had been protected from the cold, would not even supply fuel for his lungs. His lungs would of course act powerfully. This is all that would keep him from freezing. But it would require a great deal of fuel to keep the fire burning within him. The food would all be used up for this purpose; and then drafts would be made upon the different parts of his body. First the fat would go to the lungs, and be breathed away in the forms of carbonic acid and water. Then other parts, in proportion as they contain carbon, would become fuel for the lungs, until there would be little but skin and bones left.

202. This would be an extreme case; but something

resembling it takes place as often as cattle in our cold winter weather are turned into an open meadow and fed from a hay-stack. The hay, in such a case, is used up as fuel for the lungs; and this not being sufficient in very cold weather, a part of the body of each suffering animal is consumed to keep the remnant from freezing. Misery to the animal and poverty to the owner is the result.

203. To return to our illustration from a handful of corn given to an animal—we see what becomes of the starch—it goes to support respiration—keeps the fire burning; or, in other words, is breathed away in carbonic acid and watery vapor, supplying animal heat by its transformation. The gluten forms muscle, tendon, cartilage, and other similar parts. The oil lays fat over and among the other portions of the body. And the phosphate of lime forms the hard, solid part of the bones, constituting the framework of the whole.

204. From what has now been said it is manifest that the farmer, who wishes to expend his crops in the most profitable manner, must look especially at two things: 1st. The object for which he feeds any particular animal, whether it be to obtain labor, as in the case of horses and working oxen; or fat meats, as when he fattens cattle, sheep, and swine; or milk, as when he feeds cows and suckling ewes; or growth only, as in case of young cattle, store-sheep, and swine. If he would feed to the best advantage, he must have a settled plan with regard to his ani-

mals, and must feed them accordingly—some for working, some for fattening, some for milking, and others for growing. These different purposes for feeding require different kinds of food. 2nd. He needs to understand the composition of his crops, in order to know on which class of animals to bestow particular crops. It is not my purpose to lay down fixed rules. If the farmer observe carefully the effect of his feeding, he will learn to feed well without such rules; if not, he will never feed well, though he have as many rules as there are in an American cook-book. General principles, however, may be of use.

205. We will suppose that the farmer has disposed of his fall feed, of his pumpkins, his apples, and other perishable matters, so as to have brought his stock up to the front of solid winter in high order. This is an important step towards carrying them through profitably. We will suppose also that he has prepared for winter; that he has provided pure water in his yard, a trough under cover from which the inmates of the yard may lap salt, or let it alone, at pleasure, separate stalls for his animals as they are to be fed for different purposes, and above all, a plenty of litter for the purpose of keeping his floors dry and warm. We will suppose further that he has good hay, second-rate hay, and very poor hay; that he has corn and oats, carrots and turnips, a few quarts of bone-dust for his milch cows, which they will reject, if their hay contain sufficient phosphate of lime, but will eat with great advantage if it do not. We must suppose also that if he fully understands his business, he has provided and

stored near his stables a quantity of ground plaster, or dried peat, or, what is better, of both, to be thrown on the stable-floors for the double purpose of increasing the value of the manure and of preventing the bad effluvia from injuring his own health, and lessening the thrift of his cattle.

206. Thus equipped, he commences the solid winter. And now what disposition shall he make of his crops? According to the best analyses I can obtain, good meadow hay contains 10 or 12 per cent. of water, 4 or 5 of starch, not less than 10 of gum and sugar, 7 or 8 of nitrogenous substances, 3 or 4 of oil, 50 of woody fibre, and 7 or 8 of inorganic matter (ash). In hay grown on richly-manured land, a most valuable ingredient of the inorganic matter is phosphate of lime, while there is but little of this in hay grown on old, worn-out lands, that produce but a sparse crop.

207. From the foregoing it will be seen that good, early-cut, well-cured, meadow hay is rich in nearly all the desirable qualities for feeding. There are the starch, gum, and sugar, for keeping the lungs active, to supply animal heat. There are the nitrogenous substances for the muscles, tendons, and cartilages. There is phosphate of lime for the bones and for milk. And there is oil enough to give considerable fattening qualities. It is then suitable for all kinds of cattle. If we could procure it in unlimited quantities, and at a small price, we should hardly want anything else for the occupants of the barn. But, limited as it is,

horses, working oxen, fattening cattle, and milch cows should have about as much as they will eat. The farmer who undertakes to winter more stock than is consistent, giving the best of hay very plentifully to all these classes of animals, diminishes thereby his profits. A little of such hay should also be given daily to dry cows and stock cattle. It tends to make them hardy, and by means of it, they are rendered capable of thriving with poorer fare for a large part of their living. In some cases it may be good economy to give a less nutritious quality of hay to horses and working oxen. Hay for these may be later-cut and coarser. The most important requisite is, that it should be clean, bright, and perfectly well cured. Horses especially should have no hay put before them from which the least smoke or dust arises when it is handled, as it often gives them a disease of the lungs, called the *heaves*.

208. Second-rate hay, such as grows on poor land, or may have been washed by rains, or have been heated in the mow, is less nourishing, and should be given to dry cows and stock cattle. When fed to them in the manger, they will eat it clean and thrive very well, if furnished daily with a little first-rate hay to keep them in heart; or if, instead of this, they be treated daily to a quart or two of Indian meal, or one or two pecks of turnips. A great fault in the expending of second quality hay is that of keeping cattle on this alone. They should have something more juicy and nutritious for a part of their living. Those who have much second-rate hay should be observant of its

effect, and should intersperse more nutritious food as often at least as is necessary to keep their cattle in a thrifty condition. The eye of an animal, his hair, his motions, his general appearance, all will indicate to the observant feeder whether he is doing well. If not, a change must be made. There is no profit in keeping miserable, pinched-up, shrivelled stock; and certainly, if the owner have a heart in him, there can be no pleasure in it.

209. Besides first and second rate hay, almost every farmer is so fortunate, or unfortunate, as to have some that is very poor. He may have cut it late among the bogs of his pasture, or in a swampy part of his mowing that he has not yet found time to redeem. This may be turned to account. If he will throw it into his yard at noon in very cold weather, his cattle will eat a large portion of it. It contains carbon, and will at least furnish fuel for their lungs—will help to keep them warm, if nothing more; and will be converted into manure.

210. A portion of oat or wheat straw, and of rye straw even, may be made of some use in the same way. Cold, clear, sunny days should be selected as the best for getting rid of the poorest feed. It is observable that cattle will work it down about in proportion as they are kept in what some farmers call "good heart," by their morning and evening meals. A steer, for instance, that has a good breakfast and supper, will contrive to get down dry straw for his lunch; while one that fares very hard night and

morning, will not, and probably cannot, swallow dry straw, nor hay that is nearly as dry.

211. According to an analysis by Prof. Norton, Indian corn contains 12 per cent. of water, 40 of starch, 6 of gum and sugar, 17 of nitrogenous substances, 9 of oil, 14 of woody fibre, and 2 of ash. The ash of corn is rich in phosphate of lime. This renders it valuable for milch cows. It will be observed that corn gives about three times as high a per cent. of oil as good hay. It would probably yield ten times more oil than the poorest quality of hay. This renders it valuable for fattening purposes. With the exception of oil-cake, an article not much used of late years in our country, though used extensively in England, corn is the most fattening feed that we have. It is to be given freely to fattening cattle and swine. If given to milch cows, its tendency is not so much to increase the quantity of milk as to improve the quality.

212. A little corn-meal fed to milch cows daily in connection with fine, early-cut hay, or, what is equally good, if it can be obtained, with bright, well-cured rowen, gives the milk great richness. One quart of such milk is worth at least three quarts of milk from cows fed on mouldy hay and slops. Whether corn should be given to horses is a matter about which opinions differ. Some teamsters prefer it to anything else. The oat, however, seems to be natural to the horse. Horses fed on clean hay and oats seem to feel

more light and nimble; and I am inclined to the belief that they wear better and will do more service in the long run than if fed on corn.

213. It is pretty generally agreed that if horses are to be fed on corn, it should be old corn; as new corn, that which is not thoroughly ripe and perfectly dry before being ground, endangers the health of this animal.

214. Corn is excellent feed for sheep. To fattening sheep it should be given freely. For store sheep, turnips, cut into thin slices, are a good substitute for corn, and are more economical.

215. Corn is the great staple for pork-making. The very best pork and lard are made from corn. Many farmers are of the opinion, that as the prices of corn and pork have been for many years past, the manufacture of corn into pork and lard does not pay. Certainly it does not, if the feeding be done at random. But it should be remembered that the manure made by fattening swine is of very great value to the farmer: in many parts of the country we can hardly dispense with it. If dry peat, black swamp muck, dried beforehand, or rich scrapings from the roadside, be thrown into the pen in such quantities that the whole will be kept only moderately moist, there is hardly an end to the manure that can be made in this way. With ten growing and two fattening swine, a cart-load, richly worth one dollar, may be made

every day in the year, at an expense of labor not exceeding fifty cents a load, including the application of it to the field in the spring.

216. Still it must be admitted, the pork and lard will not pay for the corn, unless it be given to good breeds of swine and fed out in the best manner. But if good breeds be chosen, if the shotes be reared economically till within two or three months of "killing time," if the corn-meal then be "poured into them" without stint, I believe that pork and lard making can be made a remunerating business, especially when we take into account the great value of the manure.

217. It has been thought by many that corn-meal should be slightly sour before feeding it to swine. If wet with water, it ferments. First, there is the *saccharine* fermentation; the starch turns into sugar. Then there is the *vinous* fermentation; by this the sugar turns to alcohol. Then follows the *acetic* fermentation, by which the alcohol turns to vinegar. Now it is manifest that if we let the meal pass through all these stages of fermentation, there would be a great loss, for there are few or no fattening properties in vinegar. But many believe that if the fermentation be arrested while between the saccharine and vinous stages, while yet there is much sugar and some alcohol, the food thus prepared is congenial to the swine, and that a given quantity will produce more pork and lard than if given unfermented. This looks reasonable; and it should be thoroughly tried by those who are feeding corn to swine. For all other animals corn in its un-

fermented state is best. For human beings it is certainly better in any of those numerous forms, in which it is found on our tables, than when wrought into alcohol. I say this, however, in view of the higher end of man. If one of our race were to be fattened for a cannibal market, it is possible that even he might be made to assume a more imposing magnitude, and to become perhaps more tempting to savage eyes, by first fermenting and then distilling his corn-meal.

218. I have before spoken of oats, as the most congenial, and probably in the long run the most profitable grain for horses. Professor Johnstone gives their composition as follows:—water, 16 per cent.; starch, 38; gum and sugar, 7; nitrogenous substances, 16; oil, 6; woody fibre, 15; ash, 2. It will be observed that oats contain much oil, and a large amount of the nitrogenous, or muscle-forming matter. From this we might infer that they are good for fattening animals, and also for all kinds of working animals. Such I believe to be the facts in the case, as decided by experience. Indeed, oats are good for any kind of animal on the farm; and when they bear a price not exceeding half the price of corn, I believe it is for the farmer's interest to prepare all his provender in part from this grain. The straw of oats, if cut 5 or 6 days before the grain is fully ripe, is an excellent fodder, fully equal to third-rate hay, and better than the poorest hay. If cut in a straw-cutter, and moistened, with the addition of a little corn-meal, it is an excellent food for any animals in the barn, with the exception perhaps of fattening cattle and milch cows, and of sheep.

Milch cows should have more juicy, succulent food; and fattening animals should be spared all labor, even that of chewing tough food. They should be treated to oily food; and it should be of kinds that are easily digested; and they should be kept as quiet as possible. The passage to their stall should be by an easy ascent, so as to require little exertion in going in and out. The keeper should never strike them, nor even threaten a blow, nor in any way alarm them. Some have supposed, and I think with much reason, that they are more dormant, more perfectly at ease, and will fatten faster, if their stall is somewhat dark. Darkness promotes sleep; sleep favors digestion; and the more perfect their digestion, the larger proportion of their food will be laid over and through their frames in the form of fat.

219. Rye is of about the same weight as corn; con tains more sugar and gum than corn, but less nitrogenous matters; and not more than one-third as much oil. Where growth only is the object of feeding, as with young cattle and shotes; also where labor only is desired, as with the horse; and where labor and growth are both sought in the same animal, as with young working oxen, rye may advantageously form a part of the feed, if the price be not higher than that of corn. The composition of rye would lead to the conclusion, that if given to milch cows, it would produce as much milk as corn, and possibly more, but that the milk would be of inferior quality, and would not make as much nor as good butter or cheese. Its fattening properties are small. For the purpose of making beef,

pork, mutton, tallow and lard, unless its price should happen to be very low, not more than three-fourths that of corn, which seldom or never happens, it can not be used advantageously.

220. Experiments have shown that on deep loams, carrots may be raised to great advantage. Though not less than 85 per cent. of their weight is water, yet the quantity that can be grown on an acre is such, as to leave a very large amount of solid food, after deducting 85 per cent. They are excellent for horses, if given in small quantities, and with discretion, so as to produce on the bowels a slightly relaxing effect. When given as food in part to milch cows, the tendency is to keep the animals in a healthy state, and to impart a yellowness and richness to their milk, with none of that unpleasant flavor, that comes from feeding on turnips. They are excellent also for any kind of stock cattle, when fed upon too large a proportion of coarse and not very juicy hay. Carrots, as well as turnips, potatoes, and other roots, should, if possible, be kept in cellars that are cool. The nearer they can be kept to the freezing point, and yet not freeze, the better; for if too warm they sprout and become dry and pithy, which greatly diminishes their value.

221. Turnips contain from 85 to 90 per cent. of water; 7 or 8 of pectine (a substance similar to starch); about 2 of gum and sugar; from 1 to 2 of nitrogenous matter; less than 1 of oil; about 2 of woody fibre; and 1 of ash. The turnip is less nutritious than the carrot. It cannot so advantageously as

the carrot be given to milch cows, because it imparts an unpleasant flavor to the milk. But it has the advantage of most if not of all other crops in the great amount that can be grown on an acre, without injuring the land, but rather benefiting it, for other crops, particularly for the wheat crop. In countries that are both wheat-growing and wool-growing, the turnip culture has proved highly advantageous. The sheep eat the turnips; and the turnips and sheep prepare the ground for wheat. England can this moment raise turnips enough to feed millions of sheep, and yet raise more wheat, than if no turnips were grown. Whether such a state of things will ever be introduced into our country, time must show. But as many are doing a little at the turnip culture, it is desirable to ascertain what is the best use for these roots. They seem to possess very little of the fattening properties. For stock cattle, if turned off with poorish hay, they could not fail to be of great use, both for the substantial nourishment they contain, and from their influence in keeping the animals in a healthy condition. But their great use is for sheep. Sheep will do well on about half the hay they would otherwise require, if they can have as many turnips as they will eat. Our climate is less favorable to the turnip culture than the more humid climate of England. There, from 25 to 30 tons of turnips can be grown to the acre with *great certainty.* Our dry climate must ever render this crop less *certain.*

222. I have not spoken of potatoes, as a food for animals, because this crop has become so doubtful, owing to the blight, that if hereafter we can obtain

enough for the table, it will be about as much as we can now expect. Whenever potatoes are used for cattle or swine, there is no doubt that their value is greatly increased by cooking.

223. The same is true also of apples. Raw apples, in small quantities, are good for nearly all animals; but if cooked they are far better. Indeed, nearly all kinds of food are better for being cooked. That Indian meal for swine is worth far more when cooked, there can be no doubt. It is not so, however, with regard to all animals. Horses do better on raw food. All kinds of horned cattle and swine are more benefited by an equal amount of apples, potatoes, pumpkins, and meal of every kind, when cooked, than when raw.

224. With regard to hay, especially long, coarse hay, and all kinds of straw, the value is increased by cutting, more than enough to compensate for the extra labor. Animals more easily digest food that is properly prepared for them. The food is more readily and in a larger proportion converted into the parts of their bodies.

225. With regard to wintering stock, it has already been laid down as wretched policy to allow them to become lean and puny on the threshold of winter. The cost of wintering is thereby increased, and the profit is diminished. It is not as bad policy, yet it is not good, to allow them to fall off on the heels of winter. Some farmers get their stock along into

April in good order, and then allow them to lose half the benefit of a good wintering for the want of a little more feed and attention. They should persevere in well-doing a little longer, that their cattle may commence the summer in good plight. In this way they will make a good summer's growth; and many of them will turn for early beef, when beef is almost always higher than in autumn.

226. It is a very great error to suppose that young cattle may be turned off with the most innutritious food, and with little care and no shelter. On the contrary, they should be warmly sheltered, kindly cared for, and fed with nutritious food. The milk of the mother contains all that they need in the earliest stage of life. When weaned from this, they should be so fed as to make no unnecessarily great change in their fare. It is bad policy to give them a stint at this period, from which they will never fully recover.

227. All animals are subject to a constant waste of the body. Every few years, probably as often at least as once in seven years, the entire body is changed. The old particles have been removed, and new ones have taken their place. This waste and renewal are more rapid in young, than in older animals. From their food, therefore, must be supplied the material not only for their growth, but for that waste of the body which is so rapid at this age.

MILK.

228. It is well known that the milk of some cows is

rich, and that of others poor; and that the calves of the former will fatten rapidly, and those of the latter remain lean. The milk of some is excellent for butter; of others for cheese; and of others for both butter and cheese; while that of many is of little use for any purpose.

229. If you were to set in a shallow pan 100 ounces ($6\frac{1}{4}$ lbs.) of milk of a medium quality, there would arise to the surface three ounces of oily matter (cream); if you were then to take off the cream, and put into the milk a little rennet, there would be separated and suspended near the surface about 4 ounces of caseine (curd); if now you remove the curd, and evaporate the whey by a gentle heat, there will remain on the bottom of the pan about 4 ounces of a peculiar kind of sugar (lactic sugar, sugar of milk); and if you now burn the cream, curd and sugar, there will remain about half an ounce of ash. The composition of this milk then would be: water, $88\frac{1}{2}$ per cent.; cream, 3 per cent.; curd, 4 per cent.; sugar of milk, 4 per cent.; and inorganic matter (ash), $\frac{1}{2}$ of 1 per cent.

230. Other samples of milk might give a little different proportions. Such as are remarkably good for butter might possibly give as high as 5 per cent. of cream; such as are peculiarly excellent for cheese might give as high as 5 per cent. of curd; and those of the very best qualities for both cheese and butter might give as high as 5 per cent. of cream, 5 of curd, and 4 or 5 of sugar, leaving but 86 or 7 per cent. for water. As rich a sample as this would probably

give $\frac{3}{4}$ of one per cent. of ash, or inorganic matter, if burnt.

231. The ash of milk consists of phosphate of lime, phosphate of magnesia, chloride of potassium, chloride of sodium and soda. Of these inorganic or mineral substances, about one-half is phosphate of lime. Hence the importance that cows should be supplied with this substance, for they can give us nothing in their milk which they do not receive in their food. Oil-cake, corn-meal, hay from well-manured lands, and grass on rich pastures, contain sufficient of it for their purpose; while sour water-grass, and the grass on worn-out lands, is deficient in it. Cows kept on such feed, as before stated, should be supplied with phosphate of lime (bone-dust). If it can be procured in no other way, bones may be calcined (burnt in the fire till they readily fall into powder), and given to them, as they are fed with salt.

232. It will be noticed that another of the inorganic substances in milk is chloride of potassium. Some farmers have adopted the practice of mixing ashes with the salt given to their cows. I doubt whether this is well, for it will compel them to eat the ashes, in order to get the salt. It would seem a wiser course to place the ashes in a separate trough. The instinct of the animals would be a safer guide, as to whether they need more chlorine and potassium than are contained in their food. If they do, they would lap ashes, which contain one of these elements, while their salt contains the other; but if they do not, then the

eating of ashes by a sort of compulsion, with their salt, might be injurious.

233. The reader will notice, also, that chloride of sodium (common salt) is another of the inorganic substances in milk. There can be little doubt that the withholding of salt from milch cows diminishes the value of their milk many times over what the salt is worth. If we resolve the mineral ingredients of milk into their elements, we shall have phosphoric acid, lime, magnesium, potassium, sodium, and chlorine. All of these are contained in bone-dust, ashes, and salt. Now if, instead of mixing ashes with salt, as some farmers have done, or of mixing salt and bone-dust, as others have practised, we should put the three things in separate troughs, there is reason to believe that the instinct of the animals would be an unerring guide with regard to them all. To say the least, the offer of them could do no harm in any case; in some cases it might be beneficial.

234. We must remember that the cow *creates* nothing. She *manufactures* milk out of materials contained in her food. If any one of these materials fails, the whole operation is thwarted; for those proportions which Divine skill has established cannot be essentially varied.

235. Milk is sometimes said to be a *solution* of curd, sugar, and oil, together with the above-mentioned mineral substances, in water. This is not strictly correct. Solutions are transparent. Salt, for instance, is

white; but, if dissolved in water, it does not render the water white, but leaves it transparent; whereas milk is white, opaque, not transparent, which shows that the substances combined with water to form it are not perfectly dissolved, but that a portion of them at least are only suspended in water. The same appears also from the fact that some of them separate by standing.

236. The curd of milk exists in the form of little sacks, or bags, each enclosing a globule of oil. These little sacks of oil are so nearly of the same weight with the water, that, by the slightest agitation, they are kept diffused nearly equally throughout. They are, however, a trifle lighter than the water. This gives them a tendency to the surface; and it accounts for the fact that the milk in the upper portion of the cow's udder, that which is drawn last, is the richest. It is so with milk standing in a pail only a short time —the top is richer than the average of the whole. This slight tendency of the curd-sacks, which enclose the oil, of which butter is made, to rise to the surface, is the principle on which the cream is separated.

237. Owing to the upward tendency being so very slight, milk should be set in broad, shallow pans. A given quantity, set in such pans, will produce more cream than if set in deep vessels. We must suppose that the oil is lighter than the curd-sack in which it is contained. Those sacks which contain most, rise first; those which contain less, rise more slowly; and

some contain so little that they do not rise at all. There is, therefore, no doubt that the whole milk contains more oil than the cream; but whether, in actual practice, more of the oil can be separated and made into butter by churning the whole milk, than by churning the cream only, is not so clearly decided. Theory would seem to favor the affirmative, but careful experiment only can decide.

238. It may be remarked here, that the oil contained in milk is of two kinds, which can be separated by pressure: one, a yellowish, liquid oil, called *Oleine;* the other, a white, solid substance, somewhat resembling tallow, called *Margarine.* The learner should also bear in mind that oil is a *non-nitrogenous* substance (CHO), and that curd, or *caseine*, is one of those *nitrogenous* substances before spoken of, as containing nitrogen. Now it is a general rule of chemical compounds, that those which are composed of but few elements are more permanent in their nature; and that those which are composed of many are more perishable. It is also a well-known fact, that compounds containing nitrogen, when they begin to be decomposed, become exceedingly offensive. Accordingly the oil of milk, if entirely separated from other ingredients, is very permanent, is not easily decomposed, and does not readily become offensive to the taste or smell; while curd, containing, as it does, carbon, hydrogen, oxygen, and *nitrogen*, and a little *sulphur* and *phosphorus*—no less than six elements—is most easily decomposed; and when decomposed, it becomes almost intolerably offensive, and, acting like yeast, it

communicates putrefaction to whatever it touches. These facts will appear important when we come to the subject of the next section.

BUTTER.

239. I do not propose to go into all the *mysteries* of making and preserving butter, but to give some general facts which those who are desirous of learning may turn to account. It has already, been stated that cream is a mixture of oil, or butter (for, with the exception of a little salt, it is the same thing) and curd. The butter, in small globules, is wrapped up in little sacks, or bags, of curd.

240. Now the thing to be done, in order to make butter, is, to break open these sacks, and let the butter out. When this is done, we say, "The butter comes;" and sure enough it does come—comes out of the sacks. Those globules which were before kept apart by the sacks, come together, thousands of them, to form a particle large enough to be seen by the unaided eye. And now does the reader say, the more violently the churning is done, the sooner will the sacks be broken? Not so. You cannot break them by mechanical force: it is a chemical process. Put them in the right circumstances, and they will break open of themselves. Pounding will not break them. They will slip away from under the blows unbroken, just as a foot-ball will leave your foot when you give it a hard kick, but will leave it whole. Pressure will not break them. Nothing will break them till you

put them into *the right circumstances*, as to *temperature* and *exposure to air*.

241. At 40° Fahrenheit, you might churn from January to March, or at 100°, you might churn from June to September, and no butter would come. Or if you were to exclude the air entirely from the inside of the churn, you might roll that churn, with the cream in it, from Cape Horn to Labrador, and the butter would not come.

242. All the processes of nature have their conditions. The separation of butter from curd is one of these processes. *The conditions must be complied with.* We will suppose that the cream is from cows that give good milk. The farmer is unwise who keeps any other. Some cows' milk will not give much butter, for there is not much butter in it. We will suppose also that the milk has been kept at a temperature about medium between freezing and summer heat; that the cream has been taken off while the milk was yet sweet, and has been kept in a cool place till it was a little sour, or was very near the point of souring; that it is now put into a clean churn, and brought up to a temperature of about 60° Fahrenheit, gradually and without much stirring; and that we now begin to lift the dasher, or turn the crank, as the case may be, either forcing air into the cream by some patent contrivance, or at least letting air have free access to its surface, and now let us see what happens.

243. By stirring the cream we change the surface

often, and thus bring all parts of it successively into contact with the air. The oxygen of the air combines with the curd, and renders those little sacks, into which it is formed, brittle, so that they crack open, and let out the enclosed globules of butter. These come together, forming larger masses, until, if the churning be continued long enough *to gather the butter*, as it is sometimes called, nearly the whole will be found in one mass. The curd is now nearly separated. It is floating in the buttermilk. The sugar of milk is diffused through both the buttermilk and the butter, giving a peculiar sweetness to the butter, and also to the buttermilk, if the cream had not become too sour before churning. This is an important consideration; for it is this sugar of milk that performs the double office of giving to the butter a luscious flavor, and of causing it to keep well.

244. Washed butter may have a tolerable flavor at first, for it will retain a part of the sugar of milk in spite of bad management. But it will have given up to the water too much of its sugar of milk to allow of its keeping for any considerable time. Put down a firkin of butter that has been washed, and another precisely like it in every other respect, but which has seen no water, let them be from the same churning, be put into similar firkins, and kept in the same place, and the unwashed will keep best for an absolute certainty. No more absurd practice ever came into vogue than that of washing butter in floods of water. There is some advantage in washing very rancid butter, for some of its bad properties may be washed out.

It may be made tolerable. But if we wash fresh butter, we wash away the part that is essential to its richest flavor and to its preservation. No water should be put into the churn, and none used in the process of working.

245. The butter should be taken from the churn with a wooden ladle; should be worked with the same; when nearly all the buttermilk is worked out, pure, fine salt should be added; it should be salted to the taste. More salt than is requisite to gratify the average taste for this article, has no tendency to preserve butter, but rather the reverse, unless the salt is absolutely pure, which seldom happens. Most salt contains a little lime and a little magnesia; and when this is the case, any more than enough to salt to the taste, not only gives the butter a bitter flavor, but actually hastens its putrefaction. It is very important that the best of salt, as pure as can be obtained, should be used for butter.

246. I will here lay down a rule by which the dairyman can tell whether his salt is sufficiently pure for the purpose. To eight lbs. of salt, in a clean wooden vessel, add one pint of boiling water; let it stand an hour; pour it upon a thick strainer, and let the water pass into another vessel. The lime and magnesia, if any were present, have passed through in the water, together with a part of the salt—possibly a quarter of the whole. What remains on the strainer is nearly pure salt. Let that which has fallen into the vessel be put into the cattles' trough. There need be no

waste if all the salt used in a dairy were thus washed. Now, with washed salt, let a lump of butter be salted; and let another, from the same churning, be salted with some of the same salt unwashed. If the latter have a bitter taste, from which the former is free, you may conclude that the salt contains lime, or magnesia, or more probably both; and that the whole should be washed, as above described, before being used for butter, or else its place should be supplied by purer salt.

247. Many a pasture has been blamed for producing bitter weeds, when all the bitterness was in the salt. The pasture was well enough, but the salt manufacturer could make half-purified salt cheaper than pure.

248. We have said that all the buttermilk must be worked out. This is true, but it is liable to be misunderstood. What is buttermilk? It is water, with fine particles of curd, a very little oil, and a little milk-sugar in it. The particles of curd give it a whitish appearance. Now, the butter must be worked till this whitish appearance has ceased, but not till the last drop of liquid has left it. The best butter in the world is full of fine particles of a transparent liquid. It would not be best to work these out if you could, for the butter would then become tough and waxy. More butter is damaged by not working it enough, but much is damaged by working it too much. The dairy-woman should watch the complexion of what flows from the butter as she works it. When this becomes perfectly transparent, limpid, like pure water, with not the least whitish appearance, the operation

should cease at once, for whatever is taken out after that is a damage and not a benefit to the butter. It is not buttermilk, it is water, with a little salt and sugar dissolved in it, and is an essential part of good butter.

249. I have used firkin-butter from Madison County, N. Y., nearly a year old, which was as fragrant and as sweet as new-made butter; and, on examining it with a microscope, I have found it full of exceedingly fine globules of a transparent liquid. If rubbed with a knife-blade, these would run together and form drops, as limpid as spring-water. Could they have been analyzed, I have no doubt they would have been found to contain salt, water, and sugar, but no curd. Had they contained the least curd, it would have putrefied, and would have spread putridity, offensive to taste and smell, throughout the mass.

250. I have before stated that the nitrogenous substances, *curd* (caseine), *gluten* (as the tough, stringy part of wheat-flour), and *albumen* (as in eggs), are quick to putrefy, and that they always act, as yeast, to spread putrefaction. It is on this principle that a particle of curd in butter will create and spread putrefaction all about it. The sugar of milk contained in these transparent globules of liquid is conservative; the salt dissolved in them is conservative, if it be really pure salt; but the curd, if there be any, is destructive. The true idea therefore of working out all the buttermilk is, to work out all the curd, and there to stop, and not go on and work out all the life and

flavor and conservative principle of the butter, leaving it, as some do, little else than a mass of dry wax.

251. When butter is to be preserved for future use, it should be put down in wooden firkins. Stone pots, unless glazed better than we commonly find them, are porous. The mould which gathers on the outside, works its way through to the butter. It is not so with wood. The pores fill with water, so as to become nearly impervious. Besides, pots of sufficient size cannot easily be obtained. The larger the mass of butter, the better it keeps. Whether the firkin be large or small, it should, if possible, be filled at once. If this cannot be done, the top of the old should be taken off, and the staves of the firkin thoroughly cleansed, before adding new. We all know that the surface of butter, when it comes in contact with the firkin, very soon begins to putrefy. Something foul gathers along the edge, where the air, butter and wood all come in contact. A sort of rancidity commences there almost at once. If this is not taken off, it will communicate itself to the whole mass. Some cover the top with brine, but this only makes bad worse. The whole should be kept as dry as it can be in an ordinary cellar.

252. When new butter is to be added to a tub partly filled, the staves, after removing the surface-butter for an inch at least, may be cleansed by scraping the butter from them and then rubbing them with a cloth moistened in a weak solution of saltpetre, carefully sponging off with a dry cloth any water which may have fallen on the butter. The new should be

put on immediately, and the tub covered so as to exclude the air as much as possible; and it should be opened only as often as is necessary to make additions. Let no salt be put between the layers. The whole should be incorporated in one solid mass, as impervious to the air as possible. No brine should be put on the top; the tub should be filled to the brim; and then it should be kept in a dry and cool place. I know it is troublesome to put down butter so that it will keep for a certainty, but *it can be done.* Those who prefer to eat stinking butter, or to offer it in market, can avoid the trouble. I am only showing that those farmers who prefer sweet butter *can* always have it, *if they will.*

253. May, June, and October are the best months in our climate for packing butter, but with great care it can be safely done through the whole summer. If on the last working a very little sugar be added, not more than one ounce to 5 pounds of butter, it will keep good with greater certainty; and for most tastes the flavor will not be injured, but for many will be improved. The sugar should be of the purest kind. It would not do to trust to the pulverized sugar of the stores. That might contain impurities which would injure instead of preserving the butter. The sugar should be the best double-refined lump-sugar; and it should be pulverized very finely, and worked evenly through the mass. With this addition of sugar, butter may be pretty well preserved without all the care and trouble spoken of above.

254. For a succession of years, I have seen store-butter, of not much more than a medium quality, selected and put down in June; and yet in every case, when not put in stone pots, it has turned out in the following winter such that no gentleman would be afraid to eat it, nor ashamed to offer it to his friends, nor would be willing to deduct more than one cent a pound, if he were to carry it to market, from the highest price of fresh butter.

255. The cellar in which these experiments have been made is spacious, airy, cooler than most cellars, but rather damp. Its dampness may have been the reason of the failure in every attempt to preserve butter in earthen pots, while every trial with wood firkins has succeeded admirably, the butter in every trial, not less than ten in all, coming out seemingly quite as good in the winter as it went in the preceding summer. I will therefore state, that, in a dryer atmosphere, possibly stone pots may answer a better purpose than I have laid down. My own experience, whether in preserving butter at home, or in buying that preserved by others, has, in every instance, been against the use of stone pots.

CHEESE.

256. It has been stated that about 4 per cent. of milk is sugar. Now if milk be kept some time in a warm place, the caseine, or curd, acts upon the sugar, and changes a portion of it into a peculiar acid, called *lactic acid*.

257. It will be recollected that soda is one of the substances mentioned in a former section as contained in milk. It is the office of the soda to hold the curd in solution, a sort of imperfect solution, as before explained. Curd is not dissolved in pure water, but if a little soda be added, the curd will to some extent dissolve in it.

258. It is so with milk; it contains a little soda, in a free state, that is, uncombined with any other substance. The lactic acid, formed from sugar of milk as before explained, combines with the soda, and neutralizes its alkaline power, upon which the curd immediately appears in the form of curdled milk. This, if pressed, forms a kind of cheese.

259. The milk then has in itself all that is absolutely necessary to make cheese. This, however, would be a slow, inconvenient process, and would not result in the production of a good quality of cheese. The use of some other acid than that naturally generated in the milk, is therefore resorted to. It may be muriatic, or any of the mineral acids; or it may be a vegetable acid, as vinegar. The object of the acid is to neutralize the soda, to strip it of its alkaline property, and thus to withdraw it from its wonted office of holding the curd in a kind of solution.

260. I have said that almost any acid will answer this purpose. A kind of animal acid, called *rennet*, taken from the stomachs of suckling calves, is more commonly used. While the calf was living, the office

of this acid was to curdle the milk taken from the cow and to thus render it easier of digestion; and after he is killed, it is made to perform the same office. The stomach is preserved in a little salt and dried; and then, when wanted for cheese-making, is steeped in water; and this water is used to neutralize the soda in the milk, in order to separate the curd. If the cream is first taken from the milk, it makes what is called skim-milk cheese, which, if well made, is a wholesome article of food; and would be far better, if one were to live upon cheese mainly, than new-milk cheese.

261. In some parts of England richer cheese than new or whole-milk cheese, is manufactured. This is made by adding the cream of the night's milk to the morning's milk, and is nearly twice as rich in butter as ordinary new-milk cheese.

262. Thus it will be seen that cheese varies, with respect to the butter it contains, from nearly twice the natural quantity in the milk down to almost none.

263. Professor Johnstone, in his lectures, gives the following analyses of four kinds of cheese:

In 100 lbs.	No 1.	No 2.	No 3.	No 4.
Water, - -	43.82	35.81	38.58	38.46
Caseine, - -	45.04	37.96	25.00	25.87
Butter, - -	5.98	21.97	30.11	31.86
Ash, - -	5.18	4.25	6.29	3.81

264. From this table it will be seen that some kinds of cheese have less than 6 per cent. of butter, and

others more than 30 per cent. What will surprise most readers is, that cheese should contain so large proportions of water as the above table shows.

265. The ash of cheese, varying, as the foregoing analyses show, from less than 4 to upwards of 6 per cent., is more than half phosphates. For each cow kept on a pasture through the summer, there is carried off, in veal, butter and cheese, not less than 50 lbs. of phosphate of lime (bone earth), on an average. This would be 1000 lbs. for 20 cows; and it shows very clearly why old dairy pastures become so exhausted of this substance, that they will no longer produce those nutritious grasses, which are favorable to butter and cheese-making.

266. The temperature of milk at the time of putting in the rennet is a matter of much importance. It should be a little less than 100° Fahrenheit.

267. Special care should be taken to remove all the whey from the curd, or as nearly all as possible, before salting; and then afterwards to press the cheese thoroughly. The pressure should be more moderate at first, and then after most of the whey that remained after salting has had time to run out, the pressure should be increased.

268. The cheese should remain in the press at least two days.

269. The use of bad salt should be avoided with

the same care as in the manufacture of butter. That bitter salt, which so often finds its way into market, containing lime and magnesia, is a great enemy to success in the dairy business.

CHAPTER V.

MANURES.

RELATIONS OF SOILS TO MANURE.

270. SOME soils are so rich in all the elements of fertility, that they have not yet required manuring.

271. A few others possess such resources for a natural re-supply of the elements of fertility, as to allow us safely to predict that they never will require manure.

272. Setting aside the first—those which yet produce well *without* manure—as enjoying only a temporary exemption from the general rule, we may distribute lands, according to their relations to manure, into three classes: those which will produce well *without* manure, those which will produce good crops *with* manure, and those which will not give remunerating crops either *with* or *without* manure

273. To the first class—those producing without manure—belong: 1st, lands lying on the borders of

streams, and enriched by their overflow; 2d, lands enriched, as sometimes, but rarely happens, by mineral waters flowing upon them from adjacent lands during the winter and spring; 3d, lands in which there is much fertilizing matter yet undecomposed, but in which decomposition is constantly going on, so as to keep pace with the wants of crops. Whoever is so fortunate as to own lands fertilized in either of these ways, may, contrary to the general rule, take from them without giving to them. *There are few such lands.*

274. To the second class of lands—those producing well *with* manure—belong at least 9-10ths of all the land in the world. The owners hold it on the simple condition, that they are to put on as much as they take off. They must furnish the raw material, out of which their crops are to be manufactured, or they can have no crops. They have indeed some choice, in what form the raw material shall be supplied, and in what crops it shall be returned. They may supply it in the form of manure worth one mill a pound, and receive it back in the form of wheat worth two cents a pound; or they may supply it in the concentrated form of guano, worth two cents a pound, and receive several pounds of wheat for one of guano; but so far as the mineral ingredients of soils and crops are concerned, they are to put on what they take off, and as much of it. There is no choice here. With the exception of a few favored soils before described, this is the *immutable law* of farming.

275. To the third class of lands—those that by no

treatment will give remunerating returns—belong drifting sands, naked rocks, and marshes so situated as to preclude the feasibility of draining. That coming ages may reclaim vast extents, which now appear worthless, is possible. The wants of our race, however great they may become, will be supplied. Our present business is with that great class of lands, which are held on the condition that they will return just about in proportion as they receive.

RELATIONS OF CROPS TO MANURE.

276. Below are analyses of three soils, by Professor Johnstone—one fertile without manure; another fertile with manure; and a third hopelessly barren.

TABLE IV.

SOILS.

	Fertile without Manure.	Fertile with Manure.	Barren.
Organic Matter, ...	9.70	6.00	4.00
Silica, ...	64.80	83.30	77.80
Alumina, ...	5.70	5.10	9.10
Lime, ...	5.90	1.80	.40
Magnesia, ...	.85	.80	.10
Oxides of Iron, ...	6.10	3.00	8.10
Oxide of Manganese, . .	.10	.30	.05
Potash, ...	.20	trace.	trace.
Soda, ...	.40	"	"
Chlorine, ...	.20	"	"
Sulphuric Acid, . . .	.20	.08	"
Phosphoric Acid, . . .	.45	.18	"
Carbonic Acid, . . .	4.00	.45	"

277. The ingredients of all these soils are very well as far down as potash, except that the third has too much oxide of iron. Below that point, important ingredients are deficient in the second, and almost wholly wanting in the third. If now we look at Professor Johnstone's analyses of crops below, we shall see why the second soil required manure, and why the third was hopelessly barren.

TABLE V.

CROPS.

	Wheat.	Oats.	Barley.	Rye.	Indian Corn.	Beans.	Turnips.	Potatoes.
Potash and Soda,	33	26	22½	33	32½	45	51½	58
Lime,	3	6	2½	5	1½	8⅔	11¼	2
Magnesia,	12	10	7½	10½	16	6½	3	5
Oxide of Iron,	¾	½	1½	1½	¼	⅓	½	½
Phosphoric Acid,	49	44	39	48½	45	33	11¼	12½
Sulphuric Acid,	¼	10½	trace	1	3	4½	15	13¼
Chlorine,	trace	¼	"	trace	¼	1½	5½	4¼
Silica,	2	2¾	27	½	½	¾	2	4¼

278. From an inspection of these analyses, it is reasonable to infer that the first soil would produce any of those crops without manure; that the second would produce good crops, if manured with something containing potash, soda, and chlorine; and that the third would be likely to require more manure than the crop would be worth, and might therefore be abandoned as hopeless. The first contains all the

ingredients contained in the ash of plants; it contains plenty of organic matter; and it contains no one of the mineral substances, as oxide of iron or common salt, in such quantities as would be likely to prove hurtful. The second has also a large supply of organic matter; and it has all the mineral substances required for any crop, except potash, soda, and chlorine. This also is free from any hurtful excess of one or two ingredients. The amount of oxide of iron in it is more favorable even than in the first soil. When we look at the third, we find it not only destitute of those ingredients which are the most expensive to furnish, but abounding in oxides of iron to an injurious extent.

279. The owner of three such soils as the foregoing, could he be informed how they are constituted, would naturally cultivate crops of the most valuable kind on the first, as wheat, corn, clover. With regard to the second, he would look into the analyses of crops, and select for it those which contain least of those mineral matters in which the soil is deficient. After selecting a rotation for perhaps three years, he would next inquire how the wanting ingredients could be most readily supplied. If he were to resort to barn-yard manure, he would supply to the land a large amount of organic matter which this land does not need, because already well supplied with it. He would also supply several mineral substances with which the soil was before abundantly supplied; and he would furnish in this way but comparatively little of those ingredients which *are* really wanted. The use of

barn-yard manure would be a most expensive way of keeping up the fertility of such a soil.

280. Potash, soda, and chlorine are the things wanted. Unleached ashes contain 5 or 6 per cent. of potash, and about 2 per cent. of soda; and common salt contains 23 parts of soda to 36 of chlorine. Ashes and salt then contain all that is wanted. If now he sow on 10 bushels of ashes to the acre, and 2 bushels of common salt, his crops will probably be as much benefited as by a heavy dressing of yard manure. The one would cost him perhaps three dollars; the other would be worth thirty.

281. But it may be said that the cheap dressing will not answer the purpose always. Very true, it will not; for other mineral ingredients will ere long be exhausted, and the organic matter will also be exhausted by continued cropping, so that by-and-by a dressing of manure will become absolutely necessary. But, if the land in the mean time will produce heavy crops by means of the ashes and salt, those crops will beget manure in the owner's yard; he can put it on this land; and then the land will have manured itself, instead of drawing manure from other parts of the farm.

282. Should it ever become possible, through State patronage or otherwise, for farmers to obtain reliable analyses of their soils, farming would become somewhat an exact science. The farmer would know what crop to put on each field, and with what manure to

prepare the land. It would often happen that one dollar's worth of just what the land required for a particular crop, would benefit that crop as much as ten dollars worth of manure thrown on at random. And although special manures, containing just what a particular crop might require, and no more, would not permanently enrich the soil; yet, by producing good crops for the time, these crops would produce manure, and so furnish the means of enriching the land permanently.

IMPORTANCE OF MANURES.

283. Good farming always tends to better; and on no point is this more strikingly true, than in the care and application of manures. A load of manure well applied, not only produces a greater crop this year, but that extra crop produces more manure next year, and that extra manure produces a greater crop the year after, and so on indefinitely.

284. When speaking of lands, in another section, I shall have occasion to touch upon several kinds of manure, as better adapted than others to particular descriptions of land. If there should be some repetition, I should regret it the less from the fact that the subject of manures is the most important to which the farmer's attention can be drawn. If he manage this part of his concerns well, a foundation is laid for success throughout; if he fail here, he will fail throughout.

ANIMAL, VEGETABLE, MINERAL, AND MIXED MANURES.

285. It has been common to speak of manures as *animal*, *vegetable*, and *mineral*. A few, as hair, horns, hoofs, leather clippings, sweepings from the woollen factory, &c., are almost wholly of *animal* origin. Others, as decayed straw, vegetable mould from the woods, peat, swamp muck, &c., are almost wholly of *vegetable* origin. And others still, as plaster, ashes, common salt, saltpetre, lime, soda, &c., are purely of *mineral* origin; while a few, including barn-yard manures, are of a *mixed* character, partaking of an animal, vegetable, and mineral origin.

MANURES, STIMULANTS, AND AMENDERS.

286. Substances used to benefit soils and crops, have also been distinguished into *manures*, *stimulants*, and *amenders*. Those of which the principal object is to furnish food for plants, have been called *manures;* those whose main object is to bring into action other substances already in the soil, have been called *stimulants;* and those designed chiefly to change the physical condition of the soil, as when clay is put upon sand, or sand upon clay, or peat upon either, have been denominated *amenders*, their office being not so much to afford nutriment to plants, nor to stimulate the soil, as to better its physical state.

287. Unfortunately for the latter distinction, the object of an application is seldom confined to one of

the foregoing offices. For instance, we harrow stable manure into a clay loam. It furnishes the plant with eight or ten kinds of food; the salts contained in the manure act on the silicates in the soil as stimulants; and the manure itself, mingling with the heavy soil, renders it more open and porous. Or if we sow plaster upon a clover field, it performs at least two offices: it feeds the clover with sulphuric acid and lime, and it stimulates the soil, hastening the decomposition of organic matter contained in it.

ORGANIC MATTER—HOW TO ASCERTAIN ITS AMOUNT IN A SOIL.

288. By recurring to Table IV., it will be seen that among the ingredients of soils, is organic matter. This, so far as of vegetable origin, consists of oxygen, carbon and hydrogen, with a very little nitrogen. So much of it as is of animal origin contains the same elements, with a larger proportion of nitrogen, and a very little sulphur and phosphorus. This organic matter is essential to the fertility of soils. Its tendency is to increase in lands that are in grass, but to diminish in those under the plough, till it comes below the point essential to fertility.

289. The farmer may easily decide, whether a field is deficient in organic matter. He may take a handful of soil from half a dozen places; mix all together; dry it as dry as it can be made in the sun; put it on white paper and dry it in an oven at a temperature

just high enough to brown the paper slightly; then weigh out and put into an iron ladle 100 ounces; heat it to a red heat, and keep it hot till all the black color has disappeared; cool and weigh. The organic matter will have burned away. If it now weigh 99 ounces, his soil contains 1 per cent. of organic matter; if 98, 2 per cent., and so on. A soil should contain certainly as much as 2 per cent.; and it is well if it contain 2 or 3 times as much.

MODES OF RESTORING ORGANIC MATTER TO A SOIL.

290. When the organic matter has become deficient in a soil, there are three ways of restoring it: 1st. By laying it down to grass, and pasturing it for several years, till it has become thickly turfed over. 2nd. By ploughing in green crops. If not entirely exhausted, it may be ploughed deeply and sowed with rye, or oats and clover seed. Clover roots are inclined to run deeply in the ground. While the clover is growing, it draws for organic matter largely from the air; and at the same time, if there are valuable salts in the subsoil, it brings them up to furnish the mineral part of the crop. If, when fully grown, it be ploughed in, it not only supplies the soil with organic matter taken from the air, but with saline matters drawn up from the subsoil. If a large part of the clover be fed off by cattle, their droppings, being returned to the surface, will nearly repay the soil for the clover eaten; and then, if the remainder be ploughed in late in autumn, the effect is nearly the same. 3rd. By putting into the soil large

quantities of some kind of bulky manure; as manure from the yard, or better, this composted with 2 or 3 times its bulk of peat or swamp mud. If land, that has been ploughed so long as to have become deprived of its organic matter, is still to be kept under the plough, it must receive great quantities of bulky manure. It will not do in such cases to rely upon anything else.

OBJECT OF APPLYING MINERAL MANURES.

291. If we look again at Professor Johnstone's analyses of soils (Table IV.), we shall find among their constituents all the bases mentioned in Table III., and several of the acids. These do not exist in soils separately, but in combination with each other as salts. For instance, phosphoric acid and lime are found as phosphate of lime; carbonic acid and lime, as carbonate of lime; sulphuric acid and lime, as sulphate of lime; and so each of the acids may be combined with other bases, forming various salts. Chlorine and soda are more usually found in combination, as common salt.

292. When we apply mineral manures, it is for the purpose of adding these salts to the soil. It often happens that a soil containing a good supply of organic matter, and otherwise in an apparently high condition, will not produce a particular crop, because it lacks one or two mineral ingredients, which that crop requires. In the ashes of clover is found a considerable quantity of both sulphuric acid and lime; consequently that

crop cannot be grown on land destitue of these ingredients. But such land, by the addition of plaster, which is composed of sulphuric acid and lime, will produce clover abundantly.

293. Some have supposed, that, if we could ascertain precisely the wants of our crops, the labor of applying heavy, bulky manures might be avoided; that by spreading on our fields a few pounds of some mineral, we might carry off as many cart-loads of produce, and continue to do so, without further trouble. But such a course would soon exhaust the soil of its organic matter. The truth is, as confirmed by both science and experience, that if we would take off great crops, we must put on great QUANTITIES of manure. The ploughing in of green crops will do something towards keeping the land up; and the application of comparatively light, but expensive fertilizers, from abroad, may do something.

HOME RESOURCES FOR MANURES.

294. But after all, the farmer's great resource must be at home. *The farm must be made to enrich itself mainly.* Every particle of manure, made by the expenditure of crops, must be husbanded with the utmost care. Many a farmer, who has expended 50 tons of hay, and considerable grain crops, has heretofore had but 100 loads of manure, and that, too often, deprived by rains and evaporation of its best qualities; whereas he ought to have had four or five times as much, and of a better quality. I am aware that this

implies a great deal of labor, but it is the most profitable labor done on a farm.

VALUE OF MANURES.

295. The farmer's study is not to avoid labor, but to make labor *pay well;* and nothing is better established than that the labor of saving *manure*, of increasing its *quantity*, and improving its *quality*, is the most *profitable* that he can perform or employ. I will not say that manure is the farmer's *gold*, but *it is that which brings him gold.* About in proportion as the barn-cellar, the yard, and the pig-pen, are filled with manures, will the purse be filled with the shining metal; and, what is more, about in the same proportion will the farmer have the exquisite pleasure of seeing *everything on his farm* SHINE.

296. The subject of manures is the *golden* subject of agriculture. If I have written obscurely before, here I wish to write plainly. Let me talk, on this subject, not *about* the farmer, but *to* him. For the sake of being short and to the point, let me say *I* and *you*, instead of the more roundabout way of saying *the writer* and *the reader.*

297. For every load of manure, made by a sleepy, listless mode of farming, *you must make five loads.* Set this down to begin with. *Let the quality be improved.* How are these things to be done?

BARN-YARD MANURE.

298. First of all, put your barn-yard in the right shape, if it is not so already. Let it be slightly *dishing in the centre, and a little elevated at the edges.* Turn from it the eaves of the barn. Let no water run into it except what comes directly from the clouds, and, if possible, let one-fourth of this be cut off by sheds with their roofs turning outward. Above all, let no water run out of the yard, not even downward into the earth. How this last can be prevented you will soon learn. It is not by puddling nor by flagging the ground. It is of little consequence how tight or how porous the yard is, if you are only a wise man; for in that case, you will turn either fault to a good account. If the bottom of your yard be an impervious hard-pan, it will hold your manure of course; if it be an open, porous soil, what you would call *leachy*, you must lower it several inches every time you clear it of manure. In this way you will carry to your lands the salts of the manure, which would otherwise, in process of time, be washed into the earth, in spite of my advice to let nothing run from the yard, even downwards.

299. How are you to prevent water from running from the yard downwards, if the ground be porous, or from running over, in great rains, if it be impervious? Answer: You are to have great quantities of absorbent vegetable matters always in readiness on your farm. It may be of half-rotted straw, though, if you are a thriving farmer, your stock will be likely

to have eaten a good part of that, and made manure of the rest. More probably it will be peat or swamp mud, thrown up where you reclaimed swamp last year or the year before, now cured of its sourness by sun and rains, and ready for use. It may be peat which you have bought of your neighbor at 12½ cents a load, because none such is found on your own farm; or it may be loam of a good quality, which you turned up two years ago for this very purpose; or road-scrapings, which your men threw up in heaps at odd spells last summer; or vegetable mould, gathered into piles along the border of the woods. Whatever it is, we will call it a *vegetable absorbent.* It is not *litter.* Its object is not to keep cattle warm in winter, but to absorb their urine, which is worth as much as the solid excrements, or a little more, and to keep it from running to loss. All the substances just mentioned shall be called *vegetable absorbents,* in the remaining part of this work. They are supposed to contain decaying vegetable matter, some more and others less, and therefore to be valuable in themselves as fertilizers, but valuable, especially, as absorbents of rain-water and urine, and fully adequate, if used abundantly, to prevent the salts of manure from being washed away, and its gases from taking wings.

300. Before telling you precisely what to do with these *vegetable absorbents,* let me exhort you, as you wish to live and thrive by farming, to have them always at command, so that whenever your teams are not otherwise employed, you may draw them in for use. I wish also, before going farther, to explain a

most valuable property of these vegetable absorbents, which is not often thought of. They are *carbonaceous* —contain much carbon—and they are more or less clayey. Now, *carbon* and *clay* are the two things in nature best calculated to take in and hold fast everything nutritious to plants, whether gaseous or liquid. From the very day when you throw up a muck heap in your swamp, or by the way-side, it is gathering in for you the food of plants. Farmers always say the older their muck is, the better. There is a very good reason. It is gathering in. It lays the falling rain and the passing wind under contribution, and *it keeps what it gets.* If a chamber-maid should empty her *slops* upon a sand-heap, they would escape into the winds, or into the ground, or both. The sand-heap would become no richer. *It would* RETAIN *nothing.* But if she should empty them upon a pile of *carbonaceous* and *clayey* matter, such as may be found on almost every farm, they would be held fast. By repeating the process a few days, that pile would become almost as good as guano. Even if nothing were put upon it, it would become better from day to day, by what it would take from the rains and the air.

301. These *vegetable absorbents*, consisting mostly of black, carbonaceous matter, mixed with fine, clayey particles, and acting, as they do, both as *absorbents* and RETAINERS, are of very great value. It must be admitted that they are deficient in the more active salts, as compared with stable manure; but these, as will be shown hereafter, can be cheaply supplied, and then they become almost equal to the best of manures.

302. I will not object to the use of guano, poudrette, phosphate of lime, and other costly manures. I honor the men who prepare and sell them *honestly*. They are bringing into use a vast amount of fertilizing matters, which would otherwise be lost to the world—are returning to the country the phosphates and alkalies carried to the city in the shape of butter, cheese, meats, hay, and grains; and are raking open and bringing to market accumulations of birds' dung, scores of feet deep and thousands of years old. The traffic is a useful one. Farmers, who have faithfully husbanded their home resources, may find it for their interest to purchase these articles. They, of course, will best judge of their own matters.

303. But for inland farmers, those of but ordinary means, to let their muck remain untouched, and to leave the urine of their cattle to run into the ground, or to the nearest brook—things which, together, would make as good manure as guano, only not quite as condensed, and at the same time to buy foreign fertilizers at thirty, forty or fifty dollars a ton, seems to me like the height of absurdity. *Their* improvement should *begin at home*. But let us see how these vegetable absorbents should be used.

304. After removing all the manure from the yard, fill up the yard with them six or eight inches in depth. It will require a large amount of materials, and much labor; but remember it is a kind of labor that *pays*. This depth will be sufficient to absorb all the liquid manure of the yard. It will absorb also

the water of all ordinary rains, and hold it, till in fair weather it has time to evaporate, instead of running off and carrying the best of the manure with it. The benefit of its evaporation is, that when it evaporates, it goes off into the air, as pure water, or nearly so, leaving the salts dissolved in it behind; whereas, if it sinks through into the ground, it carries these salts away with it. It makes a great difference with the manure, whether the rain-water of a whole summer has left it by *evaporation*, or by *leaching*. In the latter case, it is full of active salts; in the former, its best salts, the potash and soda especially, which are easily dissolved, have been washed out of it.

305. Some practise ploughing over the contents of the yard once or twice a month during the summer. It is a much better practice to add to it as often a few loads of new material, enough at least to keep the thickness good or a little increasing, as the cows and other animals tread it down. Supposing the yard, including the portions under cover, to contain 20 square rods, which is none too large for a yard on a considerable farm, the solid manure, at 6 inches in depth, would give 56 large loads, of 50 cubic feet each; and every load would be worth more than the 3 or 4 or half dozen loads of dry, scaly stuff, that would have accumulated from the mere excrements of the animals, during the summer.

306. According to the practice of some, this should be carried out in the fall. I would by no means advise to such a course, unless you mean to put in as

much more, to lie in the yard over winter; and even then, it would be better to lay the new on the top of the old. The great thickness of the mass would protect it from being leached by winter storms: various kinds of litter would have been trodden into it during the winter; and early in the spring it would all be ready for use, unless the top might be so strawy that it would require to be thrown into heaps a few days to undergo a partial fermentation. Here would be, according to the number of cattle you had kept, and the amount of straw and coarse fodder you had thrown out, from 150 to 200 loads of excellent manure. It would have cost a great deal of labor to get in the materials; and it would be a heavy job to get it out; but in comparison with its real value, it would be at least a hundred per cent. cheaper than any manure you could buy.

370. There is an important consideration with regard to this manure, which must not be overlooked. Its value is not to be measured by its influence on the first crop. In addition to its immediate effect, it acts as a permanent *amender* of the soil. It should not be put upon peaty land. A few bushels of ashes would there do more good than a ton of it. But on almost any other soil, whether sandy, clayey, or gravelly, it essentially amends the soil for long years to come.

BARN CELLAR MANURE.

308. Every barn should have a cellar for vege-

tables, and another for manure. Both should be cool, but not sufficiently so to freeze. The vegetables should be kept but a little above the freezing point, and the manure at that point where it will undergo the most gradual fermentation possible.

309. On the bottom of the manure cellar, place from one to two feet of *peat*, if you have it; of *swamp muck*, if you have no peat; or of *rich loam*, if you have neither. I hardly need say that the cellar should be so constructed that a team can be driven through it, to dump these materials; and that a cart can be backed in at either end, to take out the manure. Have also in readiness near your stalls as much of the same material as you can *afford* to collect for a prospective return better than you get for any other labor.

310. Throw this into the stalls from day to day, enough to absorb the liquid excrements, and so mingle with the solid, as to render the whole a tolerably firm standing for the cattle. After one, two, or three days, as you find most convenient, open the scuttles and shovel the whole into the cellar below. It would be well if the stables were so arranged that the manure from horses, sheep, and cattle, should be mixed, in falling. Care should be taken that the manure do not ferment too rapidly. If it give a smell of ammonia (hartshorn), a few shovelfuls of plaster should be sprinkled over it. The temperature should be lowered by throwing open the cellar windows and doors. If this do not prevent too violent heat, water or snow may be thrown on. The fermentation should be as

slow as possible; and not the least smell of ammonia should be allowed.

311. The manure of a stable, thus preserved and gradually fermented, will be ready for use as soon as wanted in the spring, and will be from 2 to 4 times as valuable as if thrown out from windows to be frozen, thawed, and drenched, in the open air. It is painful to think how much labor has been lost, or at best has failed of an adequate reward, for the want of more labor in the right place—in the *increase*, *preservation*, and *right application* of manures.

312. This manure, if composted with peat or swamp muck in the cellar, would not be suitable for peaty or swampy lands. I do not mean that it would be of no use to such lands. Containing, as it would, all the salts and the nitrogen of the solid and liquid excrements of animals, it could not fail to be of use on any land; but since a portion of it was taken from peaty or swampy lands, it would be more effective if applied to lands of a different character. Soils are amended by the application of *unlike* rather than like soils.

313. In applying this manure, without analyzing it, to a soil that is not analyzed, we could not apply it on the principle of supplying precisely what is wanted for the intended crop; but we could apply it with a certainty that all its ingredients will either go into the first crop, or remain in the soil for future crops. The peat, or swamp muck, with which we have compost-

ed it, is a strong *retainer*. It will hold fast the gases and the salts of the animal part; whereas, if we put uncomposted manure into a light soil, the gases which it generates are liable to be blown away, and the salts to be washed away.

314. I will here state that nitrogen is considered to be among the most important ingredients of animal manures. Some have gone so far as to lay it down, that animal manures are valuable just about in proportion to the nitrogen they contain. When manures ferment, the nitrogen combines with hydrogen, one atom of the former to three of the latter forming ammonia (NH^3). This immediately combines with carbonic acid (CO^2), forming carbonate of ammonia (NH^3, CO^2), which is exceedingly volatile, and passes off into the air, where it is dissolved in watery vapor, and again returned to the earth in falling rain.

315. These facts show the benefit of sprinkling plaster on stable floors, which should always be done, and on fermenting manure heaps. The explanation is thus: plaster is sulphuric acid (SO^3), and lime (CaO), or sulphate of lime (CaO, SO^3). Now, when ammonia is escaping from manure in the form of a carbonate, if plaster is present, the ammonia and the lime exchange acids with each other, by what is called a double decomposition. The lime takes the carbonic acid from the ammonia, becoming carbonate of lime, and gives its sulphuric acid to the ammonia, making that a sulphate of ammonia, which last is a fixed, and not a volatile alkali, and therefore remains in the manure

(if not washed away by water, for it is soluble), till wanted by the growing plant.

316. Animal manures, while in course of preparation, should never be drenched with water, if it can be avoided, for then the potash contained in them, and the soda and chlorine which exist in them in the form of common salt, are dissolved and washed away. On the other hand, they should not be suffered to become entirely dry, as they sometimes will by excessive fermentation, but should be kept moderately moist. If too dry, it is difficult to keep the ammonia from escaping; and besides the loss of ammonia, there is another injurious action which takes place. Farmers generally speak of it as *burning.* They say their manure *burns.* There is more truth in this than would at once be supposed. The manure does burn, or an action similar to burning takes place. Let us see how this is. When wood burns on the fire, its carbon, about half of the whole, combines with oxygen, and passes off into the air as carbonic acid. Its oxygen and hydrogen pass off in the form of watery vapor, and nothing but a little ash is left. So when manure is suffered to become very dry, and to ferment excessively, its carbon combines with oxygen, and passes off as carbonic acid into the air; the oxygen and hydrogen pass off as watery vapor, and there is not much left. It is very nearly literal truth, to say, that "*the manure heap has burnt down.*" What remains is a little carbonaceous matter and a little ash, about the same as would have remained if it had been literally burnt in a furnace. The rest has gone into the atmo-

sphere, and may benefit the vegetation of the globe, but very little of it may fall back on the farm of the man who owned the manure. It will not do to estimate this *burnt* manure by the fact of its being black; for, according to that criterion, swamp muck, just as it comes from the ground, would be better than the richest stable manure. The truth is, *burnt manure*, however *black*, is worth but little—less than half certainly of its original value.

317. A famous instance of *burning manure* recently came under my observation. A gentleman who had come from the city "to farm it," piled up, on the south side of his barn, an immense heap of stable and yard manure, and let it lie from April till November—burnt down what was perhaps fifty loads to probably not more than twenty-five. He then, just before winter, spread it on a peat meadow, the soil of which was just about as black as the manure. For that soil, I suppose, twenty-five bushels of ashes, to be spread the next spring, would have been as good as the whole, if not better.

PIG-PEN MANURE.

318. Mythology relates that one King Augeus had stalled 30,000 cattle for many years without cleaning after them. Hercules, it is said, was appointed to the task of cleansing these "Augean stables." The wily hero, as the story has come down to us, turned a river through them, and made clean work shortly. Whether the stalls travelled with the current, we are

not informed, but the manure went down stream. Agriculturally considered, this was just about as wise as the management of some modern pig-pens.

319. I have often seen these important structures built with their roofs facing the south; the manure thrown out the south side; the eaves washing it in rainy days, and the sun scorching it in fair weather; till, between washing, and fermentation, and *burning*, there was little left. Others are so located, that rills, if not rivers, run into them, not enough perhaps to cleanse them, after the model of the aforesaid "Herculean labor," but enough to sweep away nearly all of their soluble salts. Owing to bad management, pig-manure has come into bad reputation, but it is good, nevertheless, if rightly managed.

320. The pig-pen should be so constructed that the eaves will be turned away from the manure. The ground should be in such shape that no water, except what falls directly from the heavens, can find ingress, and none find egress but by evaporation. There should be an outside enclosure, where the animals can be as filthy as their swinish nature prompts; and an inside apartment, where they can be as dry and warm as they please. If the first is not allowed them, they *may* not pay for their keeping in summer; if the last is not furnished, they certainly *will* not pay for their winter's food. *No animal can grow or fatten when suffering with the cold. It takes all his food to keep him from freezing.*

321. Let the outside enclosure be of considerable size, giving at least one square rod to the first tenant, and half as much more to each additional occupant. It is agreed on all hands that American farmers have land enough. They can afford to give their pigs a sufficient range. The ground should be dishing, the same as in the barn-yard, and for the same reason—that nothing may run over in wet weather; and the materials for the pigs to work over should be so abundant as never to evaporate to dryness in the dryest times.

322. Now, what is to be done that a lot of swine may produce, partly in the "natural way," and more by the manufacture of raw materials, ten loads each, per year, of excellent compost? If the number to be kept be ten, this would give a hundred loads. Suppose this to be the average number for the year, and let us see how the thing is to be done. In the first place, put around the outside of the pen, or outer yard, seventy-five loads of peat, swamp muck, road-scrapings, top-soil, or whatever you can best procure, and then proceed as follows.

323. After the pen has been cleared of its last year's manure, throw in plentifully of this to begin with. Let it be scattered over the whole enclosure several inches in depth. As it becomes thoroughly moistened with rains and the droppings of the animals, throw in more, and so on, through the summer and fall, throwing in, more or less, nearly as often as you feed the swine, and taking care that it always be moist, but seldom or never thoroughly drenched. The quantity

will soon become so large that it will hold the water of any ordinary rain, and withstand the evaporation of any drouth, if not very severe. If it inclines to dry up, it is well to throw over it a few quarts of plaster. Plaster is very little soluble. Five hundred lbs. of water dissolve but one lb. of plaster. It cannot, therefore, be lost by putting it on moist manure, as some other salts might be. Indeed, it should be sprinkled over all manures frequently, but especially if they incline, either in consequence of dry weather, or of too rapid fermentation, to become dry.

324. Some have supposed that the outer pen for swine should be under cover. I think not. Remember that rain does not hurt manure, unless it run through it, carrying off its soluble salts. Every drop of rain brings down ammonia and other fertilizing matters from the air. The falling rain washes the air of its impurities. After a shower, we say, "How sweet the air is." It *is* sweet, because it is *clean*. Hence, in the neighborhood of cities and large villages, and everywhere, to a limited extent, rain falls, impregnated with enriching materials. If it fall on a quantity of manure, which has sufficient depth to hold it, till evaporation takes place, it leaves these materials in the manure. Hence, the more rain the better, provided it go off by evaporation, and not by filtration. The evaporation should not go on to perfect dryness, for then the ammonia, the carbonic acid, and other gases, are inclined to escape, and the manure is approaching that state in which it may be said to be "burnt."

325. *Always moist but never leached*, should be the farmer's rule for his manures. The more manure he makes, both in his cow-yard and his pig-pen, the more easily can he keep it within this rule. A few inches of manure, spread over the yard or pen, will be dry as powder one day and thoroughly leached the next; while a depth of ten, fifteen, or twenty inches, will stand a long drouth, or hold the water of a long rain. Consequently, it generally happens to the farmer who makes manure on a liberal scale, that his manure is as much better in quality as it is more in quantity.

326. I have said, *always moist but never leached.* Closely allied to this is another rule. Who has not noticed that a pig-pen, in which the occupants are in danger of drowning, and one in which the manure is so dry as to be suffering a rapid fermentation, always smell horribly? To say nothing of the keeper and his family, the pigs themselves are less healthy in such an atmosphere, and they will thrive less on the same keeping. To keep a stinking pig-pen, is to throw away part of the feed and part of the manure at the same time. By giving corn to swine, shut up to a polluted atmosphere, the farmer loses a portion of his last year's crop; and, by letting his pig-pen "waste its sweetness on the desert air," he fails of a portion of his next year's. A valuable portion, and not a small portion, of what should produce crops next summer, is going beyond his reach.

327. *Not the least offensive odor should escape from the pig-pen* This is the rule before alluded to; and it is

as practicable as it is important. To practise it, will save something on the last year's crop; something for the next year's; something *certainly* in comfort; and, *it may be*, something in doctors' bills. In order to practise it successfully, one needs only to throw into the pig-pen, and all like places, including the vault of the necessary, plenty of peat, black mud, or top-soil even, and to see that it is always moist, but not drenched. A little plaster would be a help, but is not necessary. If it is not at hand, the other part of the prescription will suffice. Plaster, however, should always be on hand. This, and cured peat, or muck, should never be wanting about the farmer's premises.

328. The same rule should be observed with regard to every part of the premises. If others suffer bad odors about their farms, they may lose their comfort and their health; if the farmer suffers them, he will lose his *wealth* also; for these are the very quintessence of his manures; and it is a singular, but well-known fact, that growing plants absorb with avidity what is most noxious to animal life.

MANURE OF THE SHEEP-FOLD.

329. I shall not speak of this at large, because I suppose it to constitute a portion of the barn-cellar manure. If the apartment for sheep be so situated, that it cannot conveniently be thrown down with the manure of horses and cattle, then it would be well to mix peaty matter with it through the winter; and care should be taken that it do not dry up and become

hard. Let it be so managed as to be kept moist till nearly time to use it. If then composted with one bushel of plaster to the load, a peck of salt, and some additional peat, making two or more loads of compost for one of the animal excrements, it is an excellent manure for corn. If the land is in good heart, or, in case of its not being so, if 6 or 8 loads of barn-yard manure be first harrowed in, nine loads of this compost to the acre, (implying not more than four loads before composting) put into the hill while in a state of moderate fermentation, the corn to be planted immediately upon it, will secure a good crop of corn, from 50 to 90 bushels, according to the quality of land, to the acre, if the season be not peculiarly unfavorable. The peat used in composting for this purpose, should be rich, old peat, sweetened by the sun, and air, and rains, not newly dug, and of course cold and sour. If a little lime had been added to the peat, the previous autumn, it would be a valuable addition; only, care should in this case be taken not to allow the fermentation after composting to proceed too far. Let the pile be forked over promptly, if it become hot, and more peat added; for it is an important rule never to allow animal manures to ferment violently in any circumstances, but more especially not in the presence of lime, as it tends strongly to separate the ammonia, and will do so, to the great injury of the manure, if caution is not used.

330. In another place I have spoken doubtfully, perhaps unfavorably, of manuring in the hill. There seems to me to be no good reason why the manure

should all be at one point, inasmuch as the corn-roots fill the whole ground. Still, as the summers in the northern part of our country are short, it may be well to put a portion of the manure, while in a warm, fermenting state, into the hill, in order to give the corn a start. It is certainly better to give it a sudden push in this way, than to plant it so early that it will be long in coming, and then chilled and stinted after it has come. If corn must be small on the first of June, it is better that it should be small from being young, than "small of its age."

331. Sheep manure is excellent for the purpose of thus stimulating the early growth of corn. Perhaps horse and hog manure are equally good, if composted for the purpose, and applied when in gentle fermentation. These, however, must not be relied upon to hold out till the last of the season. Either the land must be in high order from previous manuring, or other manure must be harrowed in.

NIGHT-SOIL.

332. In European countries, as also in some of our cities, this has been wrought by various processes into a dry, portable, inoffensive, but very powerful manure, under the name of *poudrette*. This is one of the forms in which the fertilizing agents of the city are returned to the country, whence they came.

333. On the farm the night-soil may be put to good

use in a less troublesome way. After being carried off in the spring—or better, in the latter part of winter, while it is yet cool—the bottom of the vault should be covered, at least a foot in depth, with fine, black peat or mud, previously prepared and dried for the purpose. A little of the same should be thrown down daily through the summer, and once a week or fortnight during the winter. If a little plaster be occasionally added, it will be well, though this is not essential. The peat itself will be sufficiently *deodorizing*, if put down in such quantities as to be kept fairly moist and no more. It will withhold all foul odor. It is well to have an opening in the rear of the building, and a pile of prepared peat lying near, that it may be thrown down without much trouble, lest it be neglected. Good farming requires daily attention to many little things, and unless a previous preparation for them be made, these little things, important in the aggregate, are apt to be lost sight of. A farmer might better bring peat several miles for the foregoing purpose than not to have it. In an ordinary family, as many as five loads of a kind of poudrette can thus be made, not as concentrated nor as portable as the article bought under that name in our cities, but sufficiently so for home use, and excellent for any soils except peaty, and for any crops except it may be for potatoes and other roots. For cabbages, wheat, corn, or clover, it would be first-rate. If used for corn, and especially if used as a top dressing for old mowing, it would be well to apply plaster pretty plentifully with it. I know of nothing that will bring up red and white clover on an old mowing like it.

334. Many families make use of chloride of lime as a *deodorizer*, or *disinfecting agent*, about the privy. They pay for it ten or twelve cents a pound; and, at that, it is ineffectual unless used in considerable quantities. Peat is cheaper and better. When peat cannot by any means be obtained, black, vegetable mould from the edge of the wood, or wherever great quantities of leaves have drifted together and decayed, will answer. If this cannot be obtained, there is a sort of home-made chloride of lime, which can be prepared easily, and is worth more for agricultural purposes than it costs.

335. To prepare it, take one barrel of lime and one bushel of salt; dissolve the salt in as little water as will dissolve the whole; slack the lime with the water, putting on more water than will dry-slack it, so much that it will form a very thick paste; this will not take all the water; put on therefore a little of the remainder daily, till the lime has taken the whole. The result will be a sort of impure chloride of lime; but a very powerful deodorizer, equally good for all out-door purposes with the article bought under that name at the apothecaries, and costing not one twentieth part as much. This should be kept under a shed or some out-building. It should be kept moist, and it may be applied wherever offensive odors are generated, with the assurance that it will be effective to purify the air, and will add to the value of the manure much more than it costs. It would be well for every farmer to prepare a quantity of this, and have it always on hand.

SINK DRAININGS

336. The washings of the sink are of great value, if they can be so combined with peaty matter as to retain all the bad odors which they will otherwise emit. Where the nature of the ground will admit, it is best to run an under-ground drain from the sink, some distance, to where composting can be done, without appearing as a nuisance to the premises, though a well-managed compost heap, under the very kitchen window, would be preferable to a fetid sink. At the place selected for the purpose, let an excavation be made, large enough to contain six or eight loads of peat, swamp mud, or rich loam, with a view to enlarge it, by carrying off a load or two each year more than you put in. In the spring, after the old matter has been carried off, fill this piling full of peat, or some other absorbent, and direct the washings of the sink into it. By the end of a year the whole will have become thoroughly saturated with soap, rinsings of soiled clothes, oil, &c., &c.—matters most nutritious to plants. This, spread upon mow-land, will be quite equal to barn-yard manure, and, so far as the first crop is concerned, better. After the whole, which you put in the year before, is taken out, you may take a load or two more, by way of enlarging the excavation; and although this last may appear much like common soil, you may rely upon it to produce good grass. It is saturated with enriching materials.

COMPOSTING.

337. If a farmer proceed as I have recommended,

his composting will have already been done. As the spring opens, he will find a great quantity of manure in his yard, under his barn, in his pig-pen, under the necessary and at the sink-spout, already composted and fit for use. The work will have been done at times when the business of the farm was less driving than in April and May. The manure is fit for use this year. He is not to lie out of the use of it twelve months, as when manure is kept over for the sake of more perfect fermentation. If he wishes some of it to be warmer than he finds it, for the sake of starting early crops; or if that in the barn-yard is to be carried some distance, and he wishes to divest it of a part of its water, to make it lighter, he has but to throw it up into piles and allow it to ferment a few days. The same operation will both make it lighter to carry and warm for his seeds.

338. I have no doubt that this composting of manures at the place where they are made is the most economical and the best, as a general rule. There are three reasons for it: it preserves the manure more perfectly; it permits the principal labor to be done at odd spells, and at times when the teams can be spared for it; and it secures a gradual ripening, and a more perfect preparation of the manure at the very time when it is wanted.

339. There may, however, be exceptions to the rule. Suppose a piece of ground, designed for corn next year, to be a mile from the barn, and that the farmer's peat land lies in the same direction. He is unwilling

to lug the peat all the way home this fall, only to carry it back again next spring. Let him lay it, then, near the field where it is to be used. If it be in a part of the country where lime is known to work well on corn land (and there are few parts where it will not, if used as I am going to direct), let him mix 10 bushels of lime with as many loads of peat for each acre of his field; and let the compost, thus far prepared, lie till spring. If peat cannot be had, let him take what is most like it, as swamp mud, black mould from the edge of the wood, partially decayed leaves, mouldering turf, road-scrapings, or rich loam, if nothing better can be had. In the mean time, let him reserve from the home process of composting a few loads of rich heating manure, as that of fattening cattle, of horses or sheep. In the spring let him draw this to the field, and mix it load to load with the limed compost already there, adding for each load of the loam manure one bushel of plaster and a peck of salt. The tendency of the lime would be to hasten the fermentation too rapidly, and thus drive off the ammonia; but the plaster and salt will hold it fast, and the whole will form a compost worth more for a corn crop than 20 loads of the best stable manure, worth at least as much for the permanent good of the land, and not less than ten dollars cheaper for every acre. We have here then a process at once for cheapening the cost of production, and increasing the crop, and of thus stretching the profits at both ends. This is no speculation; it is the result of actual experiment. This very year I have seen, not for the first or second time, corn grown in the way just described, not in one instance, but in many, at a clear profit of 50 dollars an acre, on every outlay, including interest on value of land; while in other cases it has been raised in the same neighborhoods, and on equally good lands, at a cost little, if any, less than the value of the crop. The difference is too great. It shows that some, at least, "do not work it right." As our markets now are, *corn can be, and it ought to be, raised at a profit greater than attends most branches of business.*

340. In preparing compost for corn as above described, great care should be taken not to allow too violent fermentation after the barn manure is added to the limed peat. If the pile become very hot, it should be forked over, to check the fermentation and to mix the ingredients more thoroughly. If it be not forked over, care should be taken to pulverize and mix it as much as possible when throwing on the cart, and off. It is better, however, to fork it all over once or twice; and it should be applied warm, but not hot, to the soil. If the land is warm and light, it may be all harrowed in; if otherwise, it would be better to harrow in half of it, and to put the other half in the hill.

341. I cannot say that growing corn in the way just detailed would be a profitable business in every part of our country; but I *know* very well, from the closest observation and some experience, that in the part of the country with which I am most conversant, where corn is seldom worth less than 80 cents a bushel, it *can* be grown at a profit of which no farmer ought to complain. From long-continued and most careful ob-

servation, I have learned another fact—an important one in this connection—that the raising of great crops by these composted manures (cheap in everything except labor) is not a severely exhausting process to the soil. Farmers who have done it for years do not show *worse* lands than their neighbors, who have grown less profitable crops, but BETTER.

342. The remarks I have made with regard to composting in the field apply equally to the manures composted at home, as before described, except that peat need not be added. That is supposed to have been mixed in sufficient quantities beforehand. Its value would be greatly increased if, when drawn from the yard or cellar, it were composted with lime, plaster, and salt, in the proportions before named. It should, however, be with dead lime (oyster-shell, or slacked), not quick-lime; and special care should be taken to prevent a too rapid fermentation. The lime should not be added long before the whole is to be incorporated with the soil; as nothing can be more erroneous than to mix lime with animal manure and leave it any considerable time without attention; nor would it be well to compost it with manure to be used as a top-dressing.

ODDS AND ENDS.

343. It is well that there should be, somewhere in the vicinity of a farm-house, but a little removed from the sight, a *compost heap*, with materials lying always near, to enlarge it. Of the thousand things which

need to be carried off from a dwelling, in order to perfect neatness, let every one that is of any possible value as a fertilizer be thrown in this heap, and immediately covered over with the peat, or other substance used for composting this heap.

344. It would be quite surprising how fast such a heap would accumulate, and how valuable it would become in the course of a year; and the very circumstance of having such a dépôt for things to be "got rid of," would contribute not a little to the neatness and health of the premises. The peat, if that were used, would absorb the bad odors of whatever might be imbedded in it; or if that were not quite sufficient, a little plaster might occasionally be thrown over, which, together with the peat, would effectually prevent the escape of anything valuable to the compost, or poisonous to the air.

345. When the cellar is cleansed, the decaying vegetables and other matters should be thrown upon this heap. The sweepings of the garret should be disposed of in the same way. If the chip-yard and the wood-house are to be cleaned, whatever is too far decayed to be used as fuel, and not sufficiently so to be ready for the wet land, should go to the same *omnium gatherum* Any bits of spoiled meat; any brine that is to be carried out, and is not wanted for the asparagus bed; any dead animals, if not large; the hair and bristles from slaughtered swine; in short, whatever animal or vegetable matters are no longer fit for any other use, should be buried in this heap.

346. A single pound of woollen rags is worth more for the soil than the paper-maker would give for two pounds of clean linen shreds. No one would throw away the last; the first are almost always thrown away. Their value, as compared with barn-yard manure, as estimated by good judges, is as forty to one. Old boots and shoes, could they be reduced to powder, would be the very best of fertilizers; but as they cannot, and as they are slow to be decomposed, the best thing to do with them is to put them into the bottom of the holes in which trees are to be set, or under an asparagus bed, if one is to be prepared; or what is still better, they may be dug in about the roots of grape-vines. Those accumulations of scraps and parings of leather, which are seen by the shops of shoemakers and harness-makers, are valuable for the same purpose, especially for preparing the ground for grape-vines. Under an asparagus bed or a grape-vine, they act as a slow and constant feeder to the plants, lasting many years.

347. No dead animal, as a cow or a horse, should ever be drawn off and left to pollute the air. Bury it so deeply on the surface of the ground, with loam, that no effluvia will escape, and in a year the whole pile of earth thus thrown up, say 10 cart-loads, will be equal to the best barn-yard manure. If a little lime be put around the animal, and a bushel or two of ashes mixed with the earth as thrown on, the whole heap will become a great nitre-bed. Every particle of earth in the whole mass, and it may be large, will become impregnated with nitrate of lime

and nitrate of potash (saltpetre), which will render it an excellent manure.

348. Bones, consisting, as before stated, of phosphate of lime, carbonate of lime, and gelatine (glue), possess great fertilizing powers. In England, they are used very much for the turnip crop, and are regarded as an excellent means of preparing the ground for whatever crop is to succeed. There they are often ground to different degrees of fineness. If very fine, they act powerfully, but not for a long time; if coarse, their action is gradual, but very lasting. Prof. Johnstone informs us that as applied to pastures about 25 years ago, their action is still most distinctly seen; that in some cases pastures then dressed with bones, now rent for twice as much as others side by side and equally good by nature, which have had no bone-dressing.

349. Another mode in which bones are managed in England, and by some in our country, is to dissolve them in sulphuric acid. If put into a large tub, and moistened with about one-third their weight of sulphuric acid, diluted with five or six times as much water, the acid being sprinkled on a little at a time for several days, they will settle down into a salvy mass, which may be mixed with dried peat or loam, and put into the hill, or be sown broadcast and harrowed in. This is an excellent manure for turnips, Indian corn, or wheat.

350. Where few bones are to be had, as in ordinary

families, a less troublesome way of preserving and applying them is to dissolve them in moistened ashes. Take some large cask, as a sugar hogshead, set it in a cool place, a little away from any building, and out of the sun; into this, put bones enough to cover the bottom over four or five inches deep; throw upon the bones an equal quantity of strong, unleached ashes; wet the ashes with as much water as they will hold without leaching; then, from time to time, as bones accrue in the family, throw them into the cask; cover them with ashes, and wet the ashes as before. If this process be commenced in May, and continued till planting time the next year, the bones will then be ready for use, except that a few near the top will not be fully dissolved. These may be put into the bottom of the cask for the following year. The rest will have become soft, and may be shovelled out with the ashes, and with the addition of a few more ashes, in a dry state, will crumble into a powder. They have been applied, when prepared in this way, to Indian corn, several years in succession, and found to produce an excellent effect. The explanation is as follows: the alkalies of the ashes withdraw the oily part of the bones, combining with it and forming soap. The structure of the bones is thus broken up, and they are readily bruised to pieces.

351. Some have adopted the practice of burning the bones, and then bruising into a fine powder. This is the least troublesome way, but it is attended with the disadvantage that the organic portion, mostly gelatine, amounting to about one-third of the whole, is thus

lost; whereas, if they are dissolved in ashes, and *kept wet*, the organic part all remains. The cask may be left open at the top, and the falling rain will generally afford just about as much water as is wanted, but in long, dry spells more should be added; since, if the bones become dry, they not only become hard, instead of dissolving, but they emit offensive odors, and thus lose nearly all their organic part, nearly the same effect being produced upon them in this respect as by being burnt.

352. Of foreign fertilizers, as guano, bones of cattle from Central America, nitrate of soda (often called soda-saltpetre), from South America, and various others, I shall not speak in this work; nor shall I dwell on those more portable manures beginning to be prepared and sold in our own country, as poudrette, prepared from the night-soil of cities, phosphate and superphosphate of lime, made principally from the bones of animals, oyster-shell lime, and others.

353. I have already commended the enterprise of the men engaged in this business, as affording a channel through which the sources of fertility, ever flowing from the country to the city, may flow back again whence they came. The time will come when nearly all the mineral elements in the hay, grain, and roots, brought to the city, and a large portion of the organic elements, will find their way back to the country; those in a heavier form, to farms near the city; and those lighter for transportation, to farms more remote. Population will increase; there will be new facilities

for transportation; and the very sewers of Philadelphia, New York, and Boston, will empty themselves into the country, as those of London and Paris are now doing.

354. Whether the farmer can yet purchase and transport his manures from a distance with remunerating returns, is for him to decide. He should read his agricultural papers; he should be awake on the subject, and when it is proved to his sober judgment, that these manures will increase his annual profits, he should use them.

355. Till then, let him husband his home resources. On these, as what I consider his *great* if not his *only* resources, I have thought proper to dwell. In doing so, I have touched upon topics which, to the fastidious, may seem out of *good taste.* To me nothing seems in *bad taste*, or *undignified*, which can, by possibility, advance the great interest of agriculture.

356. With regard to the home means for recruiting lands, the rule is, that nothing be lost. Let but this be carried out, and our farms will be fertile. Almost every farm affords the means of increasing its own fertility, if they can only be applied. Correct procedure, in this respect, cannot fail of its reward. *The farmer who fails here*, I repeat, *will fail throughout; and the one who manages this matter rightly*, WILL SUCCEED. Heave up your peat, your swamp muck, your rich loam, if you have nothing better; have it always in readiness, improving by age; use it everywhere on

your premises without stint, as I have described, only using more if you please; and, depend upon it, you will reap far more than 6 per cent., or 12 either, on the cost of the labor.

357. It is not more certain that a snow-ball, in thawy weather, will grow by rolling down hill, than that *good* farming—*feeding the land well*—tends to *better;* and that *bad* farming—*starving the land*—tends to *worse.* The good farmer always grows a better farmer as life advances. *I have seen this out and out.* He gets a fair profit on his crops, and an additional reward in the *increasing value* of his lands. The bad farmer gets but a small profit on his crops, and loses that in the *diminished value* of his land. Poor and discouraged, why should he not grow a worse farmer? It is the very tendency of his course. It is hardly possible that he should make any other progress than from bad to worse—poor manuring, poor crops, a poor farm, and a poor man. Well, he must turn over a new leaf; and the very starting point of good farming lies in the generous husbandry and plentiful application of the home manures. This consideration, so important, as I view it, has made me unwilling to leave this subject sooner.

CHAPTER VI.

PRACTICAL AGRICULTURE.

RECAPITULATION.

358. In former portions of this work, I have dwelt somewhat upon *the chemistry of common objects*, hoping that such knowledge as I have endeavored to impart may be of some use to such as have not time to pursue the subject further. I have spoken briefly of the *geological formation of soils*, believing that the farmer, as he ploughs his fields, drains his lowlands, or looks after his herds over hill and dell, and along babbling streams, may pursue these thoughts with pleasure and profit. I have also spoken of *plants and animals*, of their relations to each other; of the latter as the consumers of the former's produce, paymasters for whatever crops he produces. Of manures, as one of the returns which animals make for their food and care, I have spoken at length, as I supposed the importance of the subject required. It remains to apply whatever of science may have been brought into notice to practical agriculture.

LAND—OWNERSHIP.

359. In most of the European countries, *land is not owned by those who work it.* The farmer, for the most part, holds his land on a lease of only a few years' continuance. A strong incentive to permanent improvement is therefore taken away; for, if the farmer makes ever so great improvements, he may not reap the benefit of them beyond the brief term of his lease.

360. Happily, it is otherwise in our country. *Here the landlord and the tenant are one and the same.* If he abuse his land for the sake of present income, he, and not another, is the loser. If he manage it with a wise reference to future productiveness, he, and not some hated landlord, is the gainer. In no country on earth is there so little apology for "skinned farms;" and yet such farms are everywhere seen.

361. About as much labor is expended as will suffice to take off what grows spontaneously. We see buildings, the wear and tear of thirty years excepted, what they were when the occupant was a young man. There are few or no permanent fences; the boulders about the premises lie where the drift agency left them; the annual produce is small, and growing less; the children, if they inherit a little enterprise from some remote ancestor, are all gone to the city or to the great West; and the farmer himself, if not preparing to go the way of all the earth, is at least preparing his farm to be left without regret.

362. In the name of common sense, why did he not double, instead of halving its value? He might have done it, and yet worked no harder, "scrimped" his family less, and been in all respects much more of a man. A little improvement each day of thirty years would have made his farm a thing to be proud of, and would have secured him a comfortable income in old age. This man failed to comprehend and to sustain the true dignity of *an American owner of land.*

363. Other farms are managed as if the owner were conscious that he is the owner of the increased value of the farm as well as of its annual products. The earth is grateful for such treatment; and the man who manages thus makes "his mark" on the world—marks the portion which falls to him with beauty and fruitfulness.

PERFECTION OF CROP-GROWING.

364. The perfection of crop-growing would be, that the farmer should know precisely what his soil contains and what his crop requires, and then apply such manures, and in such quantities, as would supply deficiencies, and no more. By less than supplying deficiencies, he diminishes the crop; by more than supplying them, he diminishes other crops, which should have taken the surplus manure; and let it be observed, that in either case he diminishes the amount of manure on that farm for all future years. It should be considered that a load of manure, well applied this year, begets a load next; that another the third; and

so on perpetually. It is on this principle that some farms which twenty years ago gave 100 loads, now give but 50; while others which then gave 100, now give 200. Good management has doubled the amount in one case, and a lack of good management has halved it in the other. In one case it has been compound interest *in;* in the other it has been compound interest *out.* The owners, with few exceptions, are to-day rich or poor accordingly.

365. This shows the importance of so applying manure that it will progressively beget its like. It shows also that the perfection of crop-growing, the thing to be aimed at, is, as above stated, to know the deficiencies of the soil, the wants of the crop, and the ingredients of manures, and to apply the manures accordingly.

PERFECTION NOT ATTAINABLE.

366. In the present state of knowledge such perfection is not attainable. Scientific investigation and practical experience are slowly, but surely, advancing our knowledge. Knowledge applied to agriculture will render attainable that which is now unattainable. At present we must proceed by such light as we now enjoy—must think it much if we can approximate what posterity will attain.

ANALYSES OF SOILS.

367. A great difference exists between an exact

analysis and what may be called an *examination* of soils. An exact chemical analysis, one that shall detect all the ingredients of a soil, and report them in their true proportions, can be made by a profound *analytical* chemist only. He must have studied profoundly, and practised with a patience that few possess. Probably there are not yet twenty men in the whole world who can do it *reliably*. An *examination* of soils is a very different thing. Almost any one can do something of this. An observing farmer can hardly walk across a field without forming an estimate of its value. His estimate will, in most cases, be very nearly correct; and let it be observed, that the better he can judge of a soil by a partial examination, the better he is prepared for his profession. The better his judgment in this respect, the less likely will he be to expend labor in vain, or without an adequate return.

THE CHEMIST ALONE CAN ANALYZE SOILS—THE FARMER CAN EXAMINE THEM.

368. The farmer should be advised, therefore, tc leave the *analysis* of soils to the chemist, assured that great good will come from it to his profession, whenever it can be done *reliably*, by *State* patronage, or at such reduced cost as *he* can afford to meet. In the mean time, he should be encouraged to *examine* soils, and to cultivate the most accurate judgment possible of their capabilities. That good judgment, which I have already ascribed to farmers, with regard to the

capability of soils, may be aided by attention to the following paragraphs.

HOW TO ESTIMATE A FARM.

369. In order to make my observations as practical as may be, I will suppose that I were about to purchase a farm. Let it be supposed to be at a fixed price, and my only question to be, can I afford to give that price?

370. In the first place, I would examine that farm, just as the plainest farmer in the country would. I would inspect the crops now on the farm. I would ascertain what had been done to make them what they are. I would inquire what amount of stock had been kept on the farm for years past; what had been the character of the stock; whether the farm is well watered; whether it has sufficient wood and fencing stuff; whether the buildings are in good condition; if not, what amount of money would make them such as would satisfy me; whether the land slopes to the south, north, east, or west, or is level; whether it is adapted to the kind of husbandry which I have most in view; how it is situated with relation to a village, to water power, and to market.

VARIETY OF SOILS—NAMES.

371. If these, and similar questions, were satisfactorily settled, I would ascertain whether the farm was

made up of *one* or *many* soils. If one kind prevailed through the whole, it might be worth while to procure an analysis, as in that case a single analysis would apply to the whole farm; whereas, if there were various kinds of land, several analyses would be required, and the expense would be greater.

372. If all the varieties of soil were found on this farm, we should have, according to a classification, recommended by Professor Johnstone, and now pretty generally adopted:—

1. Pure clay, from which no sand, or not more than 5 per cent. can be washed; containing about 60 per cent. of silica, combined with about 40 of alumina, as silicate of alumina.
2. Strong clay soil, suitable for brick, containing from 5 to 20 per cent. of silicious sand.
3. Clay loam, having from 20 to 40 per cent. of fine sand.
4. Loam, containing from 40 to 70 per cent. of sand.
5. Sandy loam, having from 70 to 90 per cent. of sand.
6. Sandy soil, having upwards of 90 per cent. of sand.
7. Peat, black vegetable matter, similar to swamp muck, except that it is filled with partly decayed roots and stems of plants.
8. Swamp muck, black, fine, similar to the last, but containing less, of partially decayed matter.

373. Soils may be distinguished according to this classification in the following manner:—Take 100 grains of soil, dried on white paper, at a temperature as high

as can be, without scorching the paper; boil it a few minutes; then, after allowing it to settle about one minute, turn off the water with the light clay suspended in it; add more water, stir it, let it settle as before, and turn off again; after repeating the operation several times, dry and weigh; what remains in the kettle is sand. Should nothing remain, or anything less than 5 grains, it belongs to the first class above, namely, *pure clay.* This, however, is seldom found, and if found, is valuable for other than agricultural purposes. If from 5 to 20 per cent. remains, it is of the second class, a strong *clay soil.* Such a soil as this would be too stiff to cultivate without amendment. It might be amended by mixing sand with it; and might itself be valuable for amending sandy soils, if such lay near it, so that the farmer could cart back and forth from one to the other. If from 20 to 40 per cent. of sand were found in the kettle, the soil would be of the third class, a *clay loam;* if from 40 to 70, a *loam;* if from 70 to 90, a *sandy loam;* if from 90 upwards, a *sandy soil. Peat* and *swamp* muck may be readily distinguished by the eye. These last cannot strictly be regarded as soils; they are collections of vegetable matter, more or less decayed; but as both are found to considerable extent, it seemed convenient to arrange them, as above, with soils.

CAPABILITIES OF A FARM.

374. If all these varieties of land were found on the farm I am speaking of, I should consider it the more valuable, because then the various parts of it would

furnish the means of amending other parts. It may be asked, why not purchase a farm which is good throughout, and needs no amendment? The answer is, that such farms are seldom found, and when found, the price is not such that every one could command them. On the other hand, there are many farms, at a comparatively low price, on which are facilities for making improvement, at a cost far less than the real value of the improvement.

375. For instance, there may be on the farm I am looking at a ten-acre slope of land, with the best possible exposure, and a good strong soil, but producing little, because turf-bound and too stony to cultivate. I may perceive that along the foot of this slope, adjoining the highway, is an old rickety, fallen-down fence; that the stones, which are now in the way of the plough, are well adapted to making a heavy, durable wall in the place of the old fence; that they would need to be removed but a few rods, and that down the hill; and this slope may be so situated, that the manuring of it from the barn-yard would be a down-hill, easy process. It might be very clear, that by running a substantial wall along the foot of this slope, 50 rods, at an expense of 2 dollars a rod, and thus using up the stones on and in the soil, I can make every acre worth 20 dollars more than is now asked for it. If so, the improvement would cost one hundred dollars, but would be worth two hundred, when made.

376. Again, there might be on this farm a five-acre deposit of swamp mud, nearly covered with water,

and producing nothing of any value. It might appear that by digging a deep ditch no great distance, the water might be drawn off and the land made exceedingly fertile. It might appear also that the mud which would be taken out, would be worth all the labor, to amend an adjoining patch of sandy soil, and that the sand might be brought with great advantage, by the returning team, to the low land. In this case, an improvement could be made at an expense far less than its worth.

377. On another part of this farm might be a deposit of pure clay, and near by a plot of sandy loam, an easy soil to work, and giving moderate crops, but not having sufficient consistency to hold manures. A few loads of clay would give it the requisite consistency. There is many a sandy loam which would be benefited more by ten loads of manure and ten of clay than by twenty of manure, because the clay enables the soil to hold the manure, whereas, if manure be applied alone, it escapes into the subsoil and into the air. This I suppose to be one of these cases, and it is evident that an amendment can be made at a cost less than its value.

378. There may be on another part of the farm a sandy loam and a clay-soil, at no great distance from each other, one not sufficiently tenacious to render it safe to commit manure to its keeping, the other a little too tenacious to be worked comfortably. It is evident that, by exchanging a few loads back and forth, the faults of both will be corrected. The clay-soil will be

made less refractory, and the sandy loam will be made capable of holding manure, and valuable amendments will have been achieved at a trifling expense.

379. In purchasing a farm, we should not look at it *merely* as it *is*, but as it *may be*. We should study its *capabilities*, see *how* they can be developed, and count the *cost*, and the probable return.

380. I have spoken of a general distribution of soils into *clays*, *clay-soils*, *clay-loams*, *loams*, *sandy-loams*, *sands*, *&c.* It remains to speak of their *physical properties*.

DENSITY, OR WEIGHT.

381. It is a singular fact, that we speak of a clayey soil as *heavy*, and of a sandy soil as *light*, meaning that the first is *difficult* to work, and the second *easy*. If we speak of them with reference to their *absolute weight*, the reverse is true—clayey soils are *light*, and sandy soils *heavy*.

382. A sandy soil weighs about 112 lbs. to the square foot; a strong clay soil, from 90 to 100; common arable land, from 80 to 90; garden mould, as it is more or less rich, from 70 to 80; and a peaty soil, from 50 to 70. Clear peat, perfectly dry, sometimes weighs as light as 30 lbs. to the square foot. In the foregoing cases the soil is supposed to be slightly moist. The denser a soil is, the longer will it retain its heat after sunset, or in a cold wind. A peaty soil

cools as much in an hour as a clay soil in an hour and twenty minutes, or a sandy soil in two hours.

FINENESS WITH WHICH SOILS ARE DIVIDED.

383. Some soils are more finely divided than others. The degrees of fineness may be compared by sifting dried soils through a coarse sieve. The finer they are the better, if their chemical composition is the same.

ADHESIVENESS OF SOILS.

384. When soils are wet, they are more adhesive than when dry; and those which are clayey are more adhesive than those which are sandy. The particles of the former adhere to each other, forming hard lumps, while those of the latter readily crumble in pieces. It follows, that of two soils, equally productive, one may be cultivated at a profit, because it can be worked at a small expense; while the other, being expensive to work, cannot be cultivated but at a less profit.

POWER OF ABSORBING MOISTURE.

385. This quality of soils may be compared by drying a quantity of different soils, and then exposing them to the air. If you dry a soil as dry as it can be made, by spreading it on a piece of sheet-iron, and holding it over boiling water, or by putting it into an oven of about the temperature of boiling water, and then exposing it to the air, it will be found gradually to increase in weight, in consequence of the water it

absorbs from the atmosphere. Peats and clays possess this power in the highest degree. The absorbing power of other soils—those neither peaty nor clayey—forms an important means of estimating their value. Sir Humphrey Davy found that 1,000 lbs. of soils of various qualities absorb in an hour as follows:

A very fertile soil from East Lothian, -	18 lbs.
A fertile soil in Somersetshire, - - -	16 "
A soil worth 45s., - - - - - -	13 "
A sandy soil worth 28s., - - - - -	11 "
Coarse sand worth 15s., - - - - -	8 "
Heath soil worth little or nothing, - -	3 "

By means of this absorption of water during the night, a portion of the moisture, which plants lose by perspiration in the day-time, is restored to them through their roots.

POWER OF CONTAINING WATER.

386. If we put different soils upon a fine strainer, previously saturated with water, and then let water fall upon them, drop by drop, till it begins to run through and fall below, we shall find that some will contain a much larger amount of water than others. According to Prof. Johnson, 106 lbs.

Of dry quartz sand will hold - - -	25 lbs.
Of calcareous (limy) sand, - - - -	29 "
Of loamy soil, - - - - - -	40 "
Of English chalk, - - - - -	45 "
Of clay loam, - - - - - -	50 "
Of pure clay, - - - - - -	70 "
Of a peaty soil, - - - - - -	still more.

It is also found that some soils retain water much more strongly than others when exposed to a dry atmosphere. Thus, if you should moisten a handful of dried sand, another handful of dried clay, and another of dried peat, with equal portions of water, and expose them to a dry atmosphere, the sand will lose its water two or three times as fast as the clay, and three or four times as fast as the peat.

CAPILLARY ATTRACTION.

387. If you thrust one end of a small glass tube into water, the water will rise inside of the tube higher than its surface on the outside. It is drawn up by the attraction of the glass, called *capillary attraction.* The same takes place in a sponge, which is but a collection of small tubes. If the lower part of the sponge touches the surface of the water, the water will be drawn upward, and will fill the whole. So if a snowball be brought into contact with water, the same will take place.

388. This *capillary attraction* exists in soils. If you fill a cup with dry soil, after having made a hole in the bottom of the cup, and then place it in a broad dish containing a little water, the water will find its way upward, till it moistens the whole soil, and appears on the surface. It is thus in the open field. Water in the subsoil is drawn upward by capillary attraction. If there is a surplus of water in the subsoil, it is drawn upward in too great quantities.

289. This should be explained. Whenever water evaporates, it carries off a great deal of heat. If a kettle of water is heated to the boiling point, 212°, it is made no hotter by fire below. Why? Because the evaporation from the surface carries off just as much heat as the fire infuses from beneath. It is so with a field, when the subsoil is full of water. The water creeps upward to the surface, and is there evaporated. At the moment of its being changed from a liquid to a vapor, it absorbs heat. This heat it steals away from the soil and the adjoining stratum of air, leaving the surface chill and cold.

390. If the sun shine upon such a soil, it may infuse a little more heat in the middle of the day than the evaporation carries off; but when the sun declines, the power of evaporation overmasters that of the sun, and the soil again becomes cold. Such lands are often the best in the world after being thoroughly drained, but till drained will produce nothing of much value.

391. A soil that is finely pulverized, permits the water to pass through it freely, whether upward or downward. The progress is downward after rains, and upward after evaporation. It may be laid down as certain, that the moisture in a cultivated soil is seldom stationary. It is always seeking, like the water in a sponge, to equalize itself throughout the mass. If you hold a saturated sponge just below a strong heat, the water in it will rise, and will nearly all escape, in the form of vapor, from the top. So it is with the soil. There falls a heavy rain. The top-soil

is more fully supplied with water than the soil below. A part of the water will slowly find its way downward, in order to equalize itself throughout. When it has attained something like an equilibrium, its tendency would be to remain nearly stationary, if there were a damp atmosphere and no sun. But if the sun shine, the air in contact with the soil becomes heated; it takes moisture from the soil; the surface becomes dry, and the water below moves upward.

RELATIONS OF SOIL TO THE ATMOSPHERE.

392. Soils not only require, in order to be productive, that the *air* should permeate them, but they have the power of absorbing from the air various gases, and of retaining them for the use of plants. Among these gases are oxygen and nitrogen, the principal constituents of the atmosphere; also, ammonia, carbonic acid, and various other gases, which are permanently or incidentally floating in the atmosphere.

393. Peaty soils have this power of absorbing nutritious gases from the air in the highest degree. Hence while peat, or swamp mud, is in process of *curing*, before being used in composts, it is continually growing better, not only by losing its coldness and sourness, while exposed to sun, air, and rain, but by the absorption of nutritious gases from the air.

394. Clay, next after peat, possesses this power in a high degree. Loams possess it in a greater or less

degree, according as they contain more or less clay; and sandy soils possess it in the lowest degree of all.

APPLICATION OF MANURES.

395. From what has now been said, it will be seen, that when we spread peat, swamp mud, or fermented manures upon our soils, we not only supply them with organic matter, but we give them that which enables them to draw more from the atmosphere for the benefit of our plants.

396. It will also be seen from the above remark that when we mix clay with a sandy soil, we not only render the soil more compact, more capable of holding water and manures, but we make it capable of absorbing nutritious gases—a power which it before lacked.

397. But suppose such a farm as I a little while ago described were now purchased. The buyer is no longer looking at it with reference to a purchase; but is solving the question how he shall manage it. Suppose it to be in April; and suppose the purchaser to be in such circumstances that it becomes necessary to make the farm produce the means for its own improvement. He cannot make them all at once. It must be a gradual operation, of many years. He finds the buildings out of repair, the fences down, the manure to be put upon the land, ploughing, sowing, planting, hoeing, haying, and summer harvesting, all just be-

fore him. Permanent repairs and all great improvements must give place for a while to the ordinary operations of growing and securing crops.

398. Among the first things to be done will be, to put the manure on the land. Here great judgment is to be exercised. We will suppose that there is a quantity of green manure about the stable windows, consisting almost wholly of the solid excrements of animals. The liquid excrements have probably run to waste. Such is yet the practice on most farms. Farmers have not learned that by losing the liquids of the barn and yard, they lose the most valuable part. We will suppose also that there is a quantity of yard-manure, consisting of the excrements of animals; peat, swamp mud, road-scrapings, brought to the yard the fall before; and such coarse hay, straw, and stalks, as may have been trodden down the past winter. If the former occupant were not a miserable farmer, he will find also a quantity of partly artificial manure, composed of say one-third excrements of animals, and two-thirds peat, swamp mud, road-scrapings, &c., together with a few ashes, and a little plaster and salt, now all composted together and fermented by a slow process into a rich, black, carbonaceous mass, quite as valuable as clear barn-yard manure. He will be likely also to find a quantity of hog-manure, a few loads of settlings about the sink, and a load or two of night-soil. These are an important part of his capital, on which to work the first summer; and if he is a wise man, he will take good care to double this part of his capital for the second year.

399. Now it is manifest that if he knew the exact deficiencies of his soils and the exact ingredients of these manures, he could appropriate them to the best possible advantage. This, however, he does not know; and in the present state of knowledge, he cannot. But it is evident, that if he has been an observing man, he can appropriate them, on the ground of an enlightened judgment, made up by experience, so that they will make him twice the return they would if thrown out at random. This last may seem to some extravagant, but it is true nevertheless. *Some farmers,* for years, *have not only made twice as much manure as others, with equal means, but have so appropriated it, as to get twice the return for the same amount, thus quadrupling the actual return for the whole.*

400. Now what shall the farmer do with these manures? We will begin with the solid excrements under the stable windows, premising, however, that there ought to have been none such, for there ought to have been mixed with the manure in the stables at least an equal amount of dried peat or something of the kind, by which all the liquid would have been absorbed, instead of running away into the ground. We might go farther, and say that there ought to have been a barn cellar, in which all the manure, solid and liquid, together with as much dried peat, mud or rich loam, should have been finely composted together, and that a little plaster should have been thrown on, from time to time, to check the too rapid fermentation and to fix the ammonia, thus bringing the manure into the right state to be used, exactly at the

right time to use it. By such management its value might have been doubled at least.

GREEN STABLE MANURE.

401. But our farmer, on his new place, has to take things as he finds them. All experience teaches that this green manure is more valuable to compost with cheaper materials than to use as it is. But he cannot do everything at first as he would, nor as he will by-and-bye. He may conclude to use half of this stable manure as it is, and to reserve the other half to compost during the summer.

402. If he were to put the half to be now used into a sandy soil, or a light loam even, valuable portions of it would escape into the air. If he put it on the surface of mow-land, there is danger that it will dry up, that too much of it will evaporate, and that the rest will be rather in the way of the scythe, than profitable to the crop. This latter mode of applying it would result well if the season should be warm and wet; but as this is always doubtful beforehand, the application would at best be too uncertain. He should rather apply it to plough-land, but to such as is clayey, or at least a heavy loam, in which case its virtues will be held in the soil; and such portions as are not exhausted by the first crop will be retained for the use of future crops. Stable manures, uncomposted, yield a large amount of nutritious gases; and there is hardly a more important principle in agriculture, than to put them into soils which have a sufficient retaining

power to hold them for the use of crops, instead of letting them escape into the air. Sandy soils and light loams are not equal to the trust. Within my own observation, 30 loads of green manure were ploughed into an acre of sandy loam in the spring of 1850. It gave 45 bushels of corn. On the same acre, 30 loads of similar manure were ploughed in, in the spring of 1852. The crop was estimated at 45 bushels, making 90 bushels both years, worth considerably less than the 60 loads of manure; and the worst part of the story is, that the land was not much amended; it would hardly produce another crop, without more manure. On similar lands I have seen better corn grown, with 7 loads of such manure, composted with twice its amount of peat, and the land essentially amended for years to come.

BARN-YARD MANURE.

403. With regard to the coarser barn-yard manure, it contains, in the substances, mixed with the excrements, that which is adapted to retain the nutritious gases of the latter. It may therefore with less waste be applied, if not too coarse, as a dressing to grass-lands, or harrowed into plough-lands. If it be thrown up into heaps, a few days beforehand, and slightly fermented, and a little plaster be added to prevent the escape of ammonia, it will be more than enough better to pay the extra expense.

COMPOST.

404. The composted manure, if he were so fortunate

as to find any left by the former occupant, would be good for almost any kinds of land. It might be used as a dressing to mow-lands, or be harrowed into lightish plough-lands, or put into hills for corn. It would be best, however, not to apply it to land of a character similar to that from which a large portion of it had been taken. If it was of peat, it would not be well to put it on a peaty soil; or, if it was made in part of swamp mud, it would be bad policy to put it back upon a swampy portion of the farm. As it consists of course largely of vegetable matter, it would do more good on a sandy or loamy soil, in which organic matter is deficient.

HOG MANURE—SINK SETTLINGS—CHIP MANURE.

405. As hog manure is known to act very quickly, and is liable to fail towards the last of the season, it would seem reasonable that it should be mixed with other kinds that operate more slowly, that the mixture might have the advantage both of acting quickly and permanently. The same remark applies to horse manure. It is better that both should be mixed with other manures.

406. The settlings about the sink are particularly rich in a few ingredients. More benefit therefore might be expected from mixing them with other manures, so that they would cover a larger space, than by concentrating them on a small patch. If chip manure should be found on the premises in large quantities, as sometimes happens, it should either be spread

on moderately wet mowing, in which there is little peat or black mud; or it may advantageously be applied to potatoes in the hill, especially if the land be not very well supplied with organic matter. In either of these cases—on wettish mow-land, or in the hills of a potato field—it will give an excellent return.

NIGHT-SOIL.

407. Night-soil should be removed to the land every spring. Its value, as a fertilizer, is greatly increased, if mixed with 6 or 8 times its bulk of dried peat or swamp mud. Its value would be still more increased, if the peat or mud, in a dry state, could have been thrown in with it daily, or once in a few days during the previous year; and this either with or without (better with) a little plaster, would have prevented the bad smell from that source, which is too often noticed about premises. *Poudrette* can be prepared in this way at little expense, and quite as effective as much that is offered in market at a high price. Night-soil is valuable for grass-land and for all kinds of grain. In whatever form it is used, it should be spread thinly over a large surface, rather than be put in large quantities in one place.

408. There is another article to which the last remark applies with great force. It is old plastering from the walls of rooms. This contains silicate of lime, carbonate of lime, hair, and what is of more value than all the rest, *nitrate of lime.* This last is a very soluble salt, and is so valuable for any of the

grain crops, but more especially for wheat, that not a particle of it should be lost. Every ounce of old plastering should be put upon the field. Even the rubbish of old brick walls should be pounded up and put upon the land. But this and old plastering should be spread thinly over a large surface. Probably a ton of either, if mixed with a compost that was to cover 5 acres, would benefit the first year's crop more than 5 tons spread on a single acre.

409. Whether the new occupant of this farm should go largely into the use of plaster is a question for him to settle on the ground. He should, at any rate, have some on hand to use about his manures. There is a strong presumption in favor of plaster on a farm upon which nothing is known of its effects by experience. He should inquire of his neighbors. If their testimony is against the use of plaster in that region, *let him not believe it*, but let him make the trial for himself. He may make it on a small scale at first, so as not to injure him much if it fails. If, on the other hand, the testimony of the neighborhood is favorable to the use of plaster, he might take it as undoubted. A hundred neighborhoods have testified falsely against the use of plaster in their particular location, to where one has over-estimated its value. Very few are the locations where plaster is not worth the purchase-money or more.

410. It is very true that plaster cannot be relied upon alone. It is not a manure in the fullest sense of the word. It contains but two ingredients, and those

are not all that plants need. Plants could not grow in plaster *alone*, but that does not prove that they should have *none*. The truth is, *it acts partly as a manure*—feeding the plants with its sulphuric acid and lime, the very ingredients which clover, corn, potatoes, and some other crops largely require—and *partly as a stimulant*—hastening, by its lime, the decay of vegetable matter in the soil. In other words, *it feeds the plants a part of their food, and it hurries the vegetable matter in the soil to feed them more.* On dry soils it performs *another important office*—that of *attracting moisture.* Some say it has not this effect. I know very well that in its unaltered state it has not. Set an open barrel of plaster in the air, and it will remain dry. But it does not long remain unaltered about the roots of plants. The sulphuric acid and the lime part company, and in their transformations they perform the three offices I have described—*feed the plants, convert half-decomposed matter into vegetable nutriment, and attract moisture from the air and from the subsoil.* This last office is important on lands that are dry. On wet lands it should not be used till they have been thoroughly drained.

411. Plaster will not do well permanently without other manure. It requires that organic matter should be present. In pastures this is supplied by the droppings of the cattle and by the decay of grass roots. On mowings it should be supplied by top-dressings; and on plough-lands by harrowing in manure. It would be as unreasonable to complain of plaster because it will not act well always without other manure, as to

find fault with roast-beef because it does not afford a suitable diet without other food. The same might be said of ashes. Land dressed with ashes alone, will soon be found in a sad condition; and yet the potash, soda, and lime they contain, are worth far more for agricultural purposes than the price generally allowed by soap-boilers. Their alkaline salts act favorably upon the silicates in the soil; they render insoluble silica *soluble*, and are therefore valuable on uplands; while on peaty lands, if well drained, and on any lands, which abound in inert vegetable matter, their value is very great.

DEEP PLOUGHING.

412. If our farmer on his new farm has disposed of his manures, provided his summer's stock of fuel, and made such repairs as are absolutely necessary in the outset, he will now find himself in the business of ploughing and getting in his seeds. The limits of this work will not allow me to follow him through his summer's career. A few things, however, I am not willing to pass in silence. One is the matter of ploughing.

413. From what was said on the subject of capillary attraction, we derive important rules with regard to ploughing. The upward and downward movement of the water extends far into the ground, if there is no impervious stratum. If there is a stratum near the surface, through which water cannot pass freely, an important process of nature favorable to vegetation is impeded. The water of excessive rains should pass

off without obstruction into the earth, and the upward flow of water, after evaporation, should be unimpeded, in order to supply the surface soil after a drouth. All who have tried deep ploughing have become satisfied that their fields are dryer for it in rainy weather, and moister in dry weather. This accords perfectly with the principles now explained. There may be soils lying on so porous a subsoil that it would be well to cultivate shallow. The farmer must look to this. In extreme cases, he may find a subsoil so open and porous that to stir it might be like knocking the bottom out, to let his top-soil fall into the earth and be lost among coarse pebbles.

414. Whenever the soil is deep and the subsoil compact, there can be no doubt that deep ploughing is greatly beneficial. If plants can have ten inches of loosened soil into which to thrust their roots for food, they are like a herd of cattle in a pasture of ten acres; while if they have but five, they are like the same herd confined to a five-acre lot.

415. On all ordinary soils, ploughing should be at least ten inches deep; and then, if the soil below that depth appears hard and compact, especially if there is anything like a shell or crust, through which water cannot pass freely, it should be stirred with the subsoil plough as much deeper. The water can then pass up and down freely. All danger from excessive rains is removed, because the water readily passes away from the roots of plants; and all danger from drouth is removed, or nearly all, because the water will freely

pass upward by capillary attraction; and it should be remembered that every particle of water which rises towards the surface, comes loaded with salts, which it brings from deep in the earth and deposits within reach of the roots of plants. Water so rising is never pure. If it enters the roots of plants, it carries salts along with it. If it evaporates, it leaves its salts behind, having brought them up no doubt in many cases from deeper in the ground than roots penetrate.

416. Thus we see that water acts not only as the drink of plants which they take in principally by their roots, but also as a carrier of food for them. It washes the air of all those impurities which would render it unfit to breathe. Falling as rain, it brings to the roots of plants, as food, whatever impurities the air contains; and then, after sinking deep in the earth, it is drawn back by capillary attraction, bringing with it such salts as it may have found and dissolved by the way.

417. The free passage of the air through the soil is almost as important as that of water. These considerations are worthy of the attentive study of the practical farmer. They teach him how to prepare his lands for crops. There must be in the soil that which the plant requires; and not only so, but it must be brought within the reach of the plant. Water and air are the plant's travelling agents. They must have free course; and to this end, the soil must be deeply mellowed. It would not be extravagant to say, that after having manured your soil the best you can, you

have not put within the reach of plant-roots all that they require; that still food is to be brought to them all the way from far above the surface o. the field to far below it, and that water and air are the carriers.

418. There is hardly a more important principle in agriculture than the one I have now endeavored to illustrate—that of deeply ploughing and finely pulverizing the soil. A caution is, however, here necessary. Suppose a field has hitherto been skimmed over to a depth of only five inches. Just at the termination of these five inches is what may be denominated the plough-floor—that stratum of earth on which the plough has always run, about as hard as a cart-path. Above this is a thin and exhausted soil. All below is hard, impenetrable by the roots of plants, and almost impervious to water.

419. If now the plough be put down to twice the depth before reached, and the whole ten inches inverted, it is manifest that the surface will be made up of soil that never saw the light before; and that the original top-soil will be buried at too great a depth. It would seem to be a safer course to lower the furrow one inch a year till the requisite depth were reached. In this case, the change would be less violent; the upper and lower soils would be perfectly mixed, and the whole would be thoroughly pulverized.

420. Nothing is better established than the benefit of mixing unlike soils; as peaty with sandy or with clayey soils; or swamp muck with any soil essentially

unlike it. Now, wherever the subsoil is different from the surface, this gradual deepening of the furrow enables us to mix soils without the labor of transportation. The farmer should carefully mark the effect. If good, he should continue the practice. If bad, he should investigate the cause. It might be owing to protoxide of iron in the subsoil. Should the subsoil be of a sickly yellow, when first turned up, but afterwards turn to a reddish brown, he might conclude that such is the case; and he might then add to the soil a little lime, or a compost containing it, and continue the process of deepening his soil; or should he deepen his furrows very gradually, this protoxide of iron would cease to be hurtful, merely by exposure to the air.

421. A deeply cultivated soil—one properly amended, if not originally good and well manured, is a laboratory in operation—*at work* for the owner's benefit. By means of the silica and alumina, its chief ingredients, it affords a safe anchorage for his plants; its salts and organic matter supply them food; and more than this, it is *at work*, drawing other food from above and below. The subsoil sends up its treasures, and the playful breezes pay it their contributions as they pass.

422. Such a soil, one perfected by diligence and skill, is in alliance with the silent and often unobserved but mighty powers of nature, for the farmer's good. It gathers from above and below for his benefit. It subsidizes the powers of nature in his behalf.

It is thus that the God of nature rewards diligence and skill; thus that He verifies his own truth, that "the hand of the diligent maketh rich."

423. Another item on which I will touch briefly, is that of *haying*. It is important that grass be cut before the seed is ripe enough to shell out, while the stalk is yet tender and juicy, and before it has changed into a tough, dry, woody fibre. Nevertheless, there are other things on a farm quite as important. The hilling of corn, before the roots fill the whole ground, is at least as important. Indeed, *it must be done then*, or never. The harvesting of wheat, rye, and oats, five or six days before the seed is fully ripe, is more important; for the grain is far better, and the straw is then valuable as a fodder, but is worth almost nothing, except for manure, if these crops are left to become fully ripe. Let the hay be cut earlier or later in July, according to its forwardness, if this can be done conveniently; but it is not so important that men should kill themselves with over-work to accomplish it, nor that the more important matters of hoeing and summer harvest should be deferred. Early cutting gives *better* hay; late cutting gives *more;* the medium time is on the whole the best; but the damage is not as great as many have estimated, if grass stands till into August.

424. I will now suppose that our farmer has done his haying and harvesting of summer crops; that before haying he made the necessary repairs on his barn and sheds; and that since haying he has made such

repairs on his house as he deems wise to make this year. He is now casting about, conscious that he has not the means of doing everything at once, and yet desirous of doing something every year for the permanent improvement of his farm. We suppose he has not made a fortune in the city to expend in fancy farming, and has no rich father-in-law to back him up if he gets into difficulty. The best he can do will be, to do one thing at a time. He would like to attack that ten-acre lot of boulders (428). But that would not help him to the means for enlarging his manure-heaps for another year. He therefore concludes, we will suppose, to commence operations on the five-acre swamp (429). He finds it surrounded with up-land except at one end, where by digging a ditch three feet deep, for 60 or 70 rods, the water might be conveyed away. We will suppose the swamp to be of an oval form, with an outlet at the southern extremity.

425. Let him go down the outlet to a point where the ditch may be commenced, having its bottom at least four feet below the general level of the bog. If more fall could be obtained, it would be better. I suppose this bog to be afflicted with so much water, that it would not do to trust to a covered drain. He decides upon an open drain through the centre, three feet deep and three wide. If possible, let this drain be straight. Supposing the whole length to be 66 rods, the cubic feet of mud to be thrown up would be 9,801, making about 200 loads of fifty feet each.

426. This should be done by the job. First let a

trial be made. Let it be ascertained how difficult the work is; what obstacles interfere; how far the work will be unhealthy, &c. Then let him, if possible, give it out by the job. There is not a man in the world who cannot do a difficult piece of work more easily by the job than by the day. Where work can be put out in this way, it is better for both parties.

427. What shall be done with this mud? In order to be washed of its sourness and sweetened by sun and air, it needs to lie where it is at least one year. If the owner can provide himself with other matter for composting in the intervening time, it is best to let it lie more than a year. For twenty years or more it will improve. But he wishes to clear his swamp, and be ready to put in side-drains; to have the water taken from every part, and the whole turned over with the plough, and sown with grass-seed. Probably, therefore, he will think best to remove this mud as soon as it becomes dry enough, and the ground becomes sufficiently hard for the feet of his cattle. It may be that this one ditch will take the water from the whole swamp. If not, which is far the more probable, then side-ditches should be cut running into this. If the nature of the ground admits, these should enter the central ditch at right angles. If a greater fall can be obtained by running them a little downwards, towards the outlet, then give them this direction. But let them, if possible, be parallel with each other, and at about equal distances. These should by all means be covered drains; should be from two to three feet deep; and if there is likely to be a large

amount of water to carry off, they should be within two or three rods of each other.

428. There will be considerable expense attending all this. But let it be remembered that five acres of the best land are to be made out of what was before an eye-sore. If this land can be made to produce two tons of good hay to the acre, annually, without much expense for manure, the owner can afford to lay out something upon it. How shall the side-drains be made? Suppose them to be cut two feet wide at the top, and the walls to slope inward, coming together at three feet in depth, in the form of the letter V; 1st. They may be filled with brush about two feet from the bottom, the brush be covered with turf, bottom upwards, and then the turf covered deeply with the mud thrown from the ditch; or, 2nd. They may be filled up about one foot with small pebbles, or broken stones, covered as before with turf inverted, and filled to the surface with the mud thrown out; or, 3rd. Tiles may be used.

429. *Brush-drains* have sometimes answered a good purpose, and have lasted many years. The coldness of the ground at such a depth prevents their decay. I do not believe, however, that the brush-drain is to be recommended. If the *stone-drain* is to be adopted, the stones should be very small, not much larger than hens' eggs, as otherwise the mice will work among them and fill them up. The amount of stone required for such a drain is large; the labor of collecting them is considerable; and, unless it be regarded as import-

ant to clear the adjoining grounds of pebbles, it could hardly be good economy to construct the stone-drain. The best of tiles, sufficiently large for these side-drains, can be purchased for a fraction over one cent a foot. If the ground is soft at the time of laying them, a piece of board should be imbedded for the ends to rest upon where they come together. They cannot fail, when properly laid, to carry off the water ; and if made of suitable clay, and thoroughly baked, they will last half a century, and even more.

430. Many lands, not considered swampy, would be greatly benefited by draining. This has been fully established by the experience of European agriculturists. Lands there, which a few years ago were not suspected of being troubled with water in the subsoil, have been drained; and their productiveness has been vastly increased. Probably there are great extents of land in our country, which are cold and sour, by reason of water in the subsoil, and which will ere long be rendered warm, light, easy to cultivate, and highly productive, by *thorough-draining*.

RECLAIMING STONY LAND.

431. Another season, when the ordinary business of crop growing ceases to press, our farmer may attack that ten-acre slope before spoken of (428). It is now covered with boulders, and is comparatively valueless. A wall is wanted along the foot of the slope, next to the highway. It is a heavy work to reclaim these ten acres, but considering their position, near the barn, it

may be made a profitable work. Let him have all things in readiness, iron-bars, a good strong stone boat, and an able pair of cattle. This will be a sufficient team, if not more than 3 or 4 men are to be employed, as nearly every stone, if the business be rightly managed, will be drawn directly down hill; and the team work will be an entirely different thing from what it would if the wall were to be at the upper edge of the slope. Two men to lay the wall, one to go with the team, and two to dig the stones and load them on the boat, would be perhaps the best force to employ.

432. Let the size and height of the wall be calculated according to the quantity of stone to be disposed of. If it were to be 5 feet above ground, from 1 to 2 below, according to the shape of the surface, 4 feet thick at the bottom and 2 at the top, the force I have described might put up just about three rods in a day, and at this rate the cost would not vary much from two dollars a rod.

PROFITS OF AMENDING LANDS.

433. It should be considered that lands of this description, having a favorable slope, are generally better, when cleared of stones, than those which are naturally feasible. There are thousands of acres in the Eastern States, which can thus be made first-rate land, at a cost, considering their nearness to market, less than their prospective value; and it is a singular fact, but one, I believe, which cannot be disputed, that the farmers in these States, of just such lands as I have

now described, and worse even, lands in many cases so stony, that instead of a wall on one side only, you would have to build a heavy wall around every 5 acres, to swallow up the stones, are this moment richer, and more intelligent, and are educating their families better, than those on our very best river lands. The truth is, these granite lands, when once reclaimed, fairly walled about, and thus cleared of stones, possess great capabilities. I know not but the prospects of the young man who commences on such lands, considering their healthfulness and their proximity to market, are as flattering as those of one who commences on the richest prairies.

MIXING SOILS.

434. On another part of the farm, to which I have directed attention, is supposed to be a bed of nearly pure clay, and near by it a sandy loam. This is no uncommon occurrence. Now the sandy loam has a little fine clayey matter in it, almost enough to make a very profitable soil to cultivate, but not quite; for although it is easy to work, yet, for the want of larger crops, it does not give a satisfactory profit. Now the probability is, that if our farmer can find a time either by carting or sledding when he can draw 15 or 20 loads to the acre of the clay, and put it upon this sandy loam, he will not receive his pay as promptly as would the man who should work for him by the day, but in the end he will receive, in the increase of his crops and in the increased value of his land, far higher wages I find almost everywhere, that the men who

have made *hard* farms good ones, are rich. I do not find that they were born rich, nor that they have married *rich wives*, but some how or other, they have *grown* rich; and I know not how to account for it, but on the supposition that this making good land out of poor, and then raising crops on it, is a pretty well-paid business. I think it is so—that *the man who makes a poor farm better, is better paid for his trouble than the one who makes a good farm poorer*. His satisfaction, if he ever reflects on his doings, must certainly be greater.

435. On another part of this farm was supposed to be a heavy clay soil, too refractory to work with remunerating results; and, side by side, as not unfrequently happens, a light sandy loam, unequal to the trust of retaining the manures committed to it. Now, if the owner should be tempted to go with his team and work for other people, at $2 a day, it may be wise; he may need the ready pay—we suppose he knows his own business;—but let him remember that a day's work with his team, in carrying back and forth, from one of these soils to the other, would be likely to bring him much more than $2 a day in the end.

ROTATION OF CROPS.

436. The prevailing system of rotation in England is what is called the Norfolk system. It is a four years' course—turnips, barley, clover, wheat, and then the same over. This is adapted to light soils—those called barley-soils. It is considered that the turnip crop, eaten off by sheep, prepares the ground for bar

ley. The clover, being sown with the barley, fills the ground with its roots, and thus prepares it for wheat. For heavier clayey soils, a six years' rotation is there preferred, in which wheat, oats, and beans, are made to occur as often as possible.

437. In this country, our climate is different. Under our scorching suns, turnips can never be grown as advantageously as in the humid atmosphere of England; and here, Indian corn, which cannot be grown there, will always be an important crop. English usage therefore throws little light on our course. That the principle of rotation in crops ought to be adopted, there can be no doubt; but, as yet, no very specific rules have been laid down, or, if laid down, they have not, so far as I am aware, been confirmed by practice. The composition of plants, so far as their inorganic elements are considered, is various. Some, it will be seen (Table V.), require a large amount of certain ingredients, while others require little of these, but draw largely upon other ingredients. We have, then, as a general rule, *to let those which are unlike in their requirements follow each other.*

438. There are other topics on which I would gladly dwell. I would gladly recall some on which I have spoken, with a view to repeat and enlarge, and to urge them on the consideration of practical farmers. But the limits I have assigned to myself are already more than reached. I cannot, however, close this little work without a few suggestions to that class of men, whom, if any, it is adapted to benefit. I have

spoken *of farming;* let me speak a few words *to farmers.*

TO FARMERS.

439. *Yours is a noble profession.* I will not be deterred from saying this, because so many have said it who were incapable of any just appreciation of what they were saying. Many have written and uttered it, who were much more willing that others should be farmers, than to be farmers themselves. It is true nevertheless. Yours *is* a noble profession.

The *merchant*, who brings manufactured goods to our door, and sells them at a reasonable profit, and thereby *lives* and enables us *to live* better than we could if we had to go all the way to the manufacturer for a gimlet, a plough, or a piece of calico, is doing well for the community. His is an honorable profession, and we are bound to honor him, so long as he pursues it honorably.

The *manufacturer*, who converts the raw material into the necessaries, comforts, and ornaments of life, and then passes them over to the merchant, to be distributed to all who want, is also doing a good work. We must honor him too, so long as he produces a good article, at a fair price. If, by a life of restless enterprise, he becomes rich, we will not envy him.

The *farmer*, who produces the raw material, and passes it on to the manufacturer, and through him to the merchant, and thence to the supply of all terrestrial wants, is at the foundation of the whole structure of human society. What a pity it would be, if some coxcomb. high up the grades of life, as *he* may vainly conceive, should look down and scorn the foundation!

I can hardly forgive the man *or woman* who speaks slightly of the intelligence, the worth, or the social importance of farmers. The farmer ignorant? It is impossible! He lives amid the communions of nature. The common mother of us all teaches him daily. The heavens always shine on him. How different with those, who, when they look around, see nothing but paving-stones, dry-goods, and hardware; and who, when they look up, see no heavens, unless they can see through brick and mortar! The works of man fill all *their* thought. What wonder if *they* fail to worship a higher God than Mammon! The *farmer* communes ever with the works of the Almighty. What should hinder *him* from being a reverent learner? He lives amid revelations. He cannot be ignorant, if he would. Away, away, ye profane ones, who speak flippantly of the farmer and his calling.

Nevertheless, it must be confessed, that farmers are not always as eager for the knowledge pertaining to their profession as would be desirable. They are not destitute of important knowledge; they cannot be; it is impossible. But their communion with the broad folio of nature, renders their habits of thought unfavorable, and sometimes averse even, to another kind of study, which, after all, they really need, in order to the highest success in their calling. The clergyman, the doctor, the lawyer, need books on their profession, and so does the farmer on his. I grant that he can learn a greater proportion of his duties without books than they, but not the whole. *The farmer needs books.* It is difficult, if not impossible, for him to reach the top of his profession without them.

I have seen with what eagerness the *merchant* runs over the prices current, and with what prying curiosity the manufacturer seeks out and appropriates the latest improvement in his line. I wish I could see the farmer as eager for the best agricultural paper, as the merchant is for the best journal of commerce, or the manufacturer for the best practical machinist. If the minister, the lawyer, and the doctor, insist upon great libraries of their professions, I wish the farmer would as resolutely insist upon a small one of his. Then would knowledge be increased; what one farmer knows all would know; and it would be a prodigious amount. It would be a kind of knowledge that is practically useful, beneficial, not to a *few*, but to the *whole world.*

CATECHISM

OF

SCIENTIFIC AND PRACTICAL AGRICULTURE.

(*Questions to be answered as below, or from the sections referred to.*)

WHAT is the *science* of agriculture? It is the *knowledge* of farming.

What is *practical* agriculture? It is the *practice* of farming.

What is the difference? The first is something *to be learned;* the second something *to be done.*

Can the *learning* be in all cases separated from the *practice?* It cannot.

If you were told to feed a horse, could you *learn* perfectly how to feed him without first putting your knowledge into *practice?* I think I could.

Let us see: 1st. You would need to know what food a horse requires; 2d. In what form he requires it, whether long or chopped, ground or whole, raw or cooked; 3d. How often he should be fed; 4th. How much at a time; and 5th. *You would want to be quick to judge by his appearance and action, whether you were feeding him in the best manner.* Could you learn all these things without some practice? All but the last.

If you were told to mow a piece of meadow, could you first learn how to mow, and then afterwards mow it? In this case I should have to unite the *learning* with the *practice.*

Is it not so with most things to be done on a farm? It is.

What two things then are essential to an accomplished farmer? That he should *know* everything that is to be done on a farm, and *be able to do it expertly.*

What would the first be called? Knowledge, or science. What the second? Skill.

Which of these is important to the hands on a farm? The last.

Which to the man who manages the farm? Both.

What does farming imply? Three things: 1st. The growing

of crops; 2d. The disposal of the crops; and 3d. The disposal of those things which are produced by the crops.

How many things are to be considered in the growing of crops? Four: the preparation of the ground; the putting in of the seed; the care of the plants till matured; and the preservation of the crop till disposed of.

How are crops to be disposed of? Partly by sale; partly as food for the farmer's family, but principally as fodder for his animals.

Why are crops to be consumed mainly on the farm? That their ingredients may be returned to the soil, to be transformed into future crops.

What are those secondary products of crops before spoken of? Beef, pork, mutton, fowls, butter, cheese, and eggs.

How are these products disposed of? Partly as food for the family; partly in barter for necessaries and luxuries not produced on the farm; and partly by sale, for the purpose of raising money.

Does the farmer raise all the animals that eat his produce, and no more? That would be impossible; for he does not know beforehand how much produce he will have; and therefore he could not know how many to raise.

If he should raise too many, what would he do? He would either sell some of his animals or buy produce.

If he should raise too few? He would either sell some of his produce or buy other animals.

Buying and selling then is an important part of the farmer's business; whom is he like in this respect? The merchant.

What kind of knowledge does he need to discharge this part of his duties well? What would be called mercantile knowledge —a knowledge of the prices current, of the present state of the market, and of the probable changes.

When the farmer manages to turn his soils and manures into crops, and these again into beef, pork, butter, and cheese, whom is he like? The manufacturer.

What kind of knowledge will best enable him to perform this part of his business? A knowledge of soils, plants, animals, and manures.

When the farmer has buildings to erect, fences to make, some implements to manufacture, and others to repair, whom is he like? The mechanic.

If then the farmer is to be a sort of a merchant, a manufacturer to some extent, and mechanic enough to be able to employ the carpenter and the blacksmith advantageously, does not his profession require great and varied knowledge? It does.

If a profession is to be estimated by the amount of knowledge required to prosecute it in the best manner, what profession is more honorable than the farmer's? None.

In farming, as in other things, there is a *best way*, and there are *inferior ways* of doing the same thing, and the profit often depends upon taking the right course: how would you ascertain the best way of doing something, as, for instance, to raise a ton of carrots? There are three ways in which I might learn it: 1st. *By experiment;* 2d. I *might be told it* by some one who knew; 3d. I might learn it from books. The first would be a slow process; for I might have to experiment ten years before I should hit upon the best course. The second and third would be very much alike; in either case I should get this piece of knowledge from another person, and it would be of little consequence whether he communicated it through the ear or the eye.

What peculiar advantage have books? This, that while we cannot command the services of a living teacher at all times, we can always command the assistance of books; and they can teach us at odd spells, as on rainy days or winter evenings.

How is the farmer to gain that extensive and varied knowledge which we have seen that his business requires? In the first place, he should be educated *for his profession* when young, as other young men are for theirs; and in the second place, he should pursue his inquiries through life—should be a *thinking*, and, to some extent, a *reading* farmer.

For explaining the reasons of things that are always occurring in life, and especially on a farm, what science is most important? Chemistry.

To what extent should a farmer undertake to learn chemistry? So far only as to enable him to understand those explanations which chemists are making for his special benefit.

What other science throws considerable light on the farmer's path? Geology.

To what other subjects should he give particular attention? To the natural history of plants; the nature, habits, instincts, wants, and capabilities of domestic animals; the use of manures; and the constitution of soils.

In application to what should he study all these things? To *Practical Agriculture.*

CHEMISTRY.

What is an *element?* 1. A *binary compound?* 2. A *ternary* compound? 3. A *quaternary* compound? 3.

What then does *binary* mean? 3. *Ternary?* 3. *Quaternary?* 3.

Give an example of an element? 4. Of a binary compound? 4. Of a ternary compound? 4. Of a quaternary compound? 4.

In how many *forms* does matter exist? 5. Give an example of a *gas?* 5. Of a *liquid?* 5. Of a *solid?* 5.

Do any bodies change their form? 6. In what circumstances

does water take the gaseous form? In what, the liquid? 6. In what, the solid? 6.

What is *chemical affinity?* 7. Of how many kinds is it? 7. What is *simple* affinity? 7. *Single elective?* 7. *Double elective?* 7.

How is a compound to be distinguished from a *mixture?* 8.

What bodies are said to be *soluble?* 9. What insoluble? 9. What is a liquid that will dissolve a body called? 9. What is a *solution?* 9. What is the great *solvent* in nature? 9.

Are there degrees of *solubility?* 10. How much quick-lime will water dissolve? 10. How much gypsum? 10. How much common salt? 10.

What is a general law of combination? 11. Explain this? 11.

What is another law of combination? 12. Will you explain this? 12.

How many elements are known? About 60. How many of these constitute essentially all objects with which we are conversant? 13. Will you give the names of those 15? 14.

What is *Oxygen?* 15. What portion of air does it constitute? 15. Of water? 15. Of all known matter? 15.

What is *Chlorine?* 16. In what form might it be supplied to soil? 16. For what crops? 16.

What is *Sulphur?* 17. What more can you say of it? 17.

What is *Phosphorus?* 18. What is it a part of? 18. How diffused? 18.

What is *carbon?* 19. Of what does it form a part? 19.

What is *silicon?* 20. What part of the solid globe does it probably form? 20. What is it in its pure state? 20. When combined with oxygen? 20.

What is *nitrogen?* 21. What part is it of the air? 21. What does it constitute with oxygen?

What is *hydrogen?* 22. How light? 22. What part of water? 22. Will it burn? 22. Does it cause other bodies to burn? 22.

What is *iron?* 23. What is said of it? 23.

What is *manganese?* 24. How found, and where? 24.

What is *potassium?* 25. What is said of it? 25.

What is *sodium?* 26. What is said of this? 26.

What is *calcium?* 27. Why are limy soils called calcareous? 27.

What is *magnesium?* 28. Of what is it the basis? 28.

What is *aluminum?* 29. Of what is this the basis? 29.

Which of the fifteen elements are *gases* when uncombined? 30. What of the other eleven? 30.

Which are *metals proper?* 31. Which are metals of *alkalies?* 32. Which of *alkaline earths?* 32.

Which are called organic elements? 33. Why? 33.

What are the letters written after the names of substances called? 35. What is their use? What does O stand for? 35. Cl? 35. K? 35. Na? 35. Fe? 35.

What do the figures after the symbols show? The *atomic weight*. See 36 and 37.

What is the atomic weight of hydrogen? 37. Of carbon? 37. Of oxygen? 37. Of magnesium? 37. Of sulphur? 37. What is the lightest of all bodies? 37 and 22.

What two elements combine to form chloric acid? (See Table I.) What is the symbol for oxygen? What for chlorine? What will be the symbol for chloric acid, if 5 atoms of oxygen combine with 1 of chlorine to form it?

What two elements combine to form sulphuric acid? (See Table I.) How many atoms of oxygen to one of sulphur? What then shall be the symbol for sulphuric acid? The *atomic* weight of oxygen being 8, and that of sulphur being 16; and 3 atoms of oxygen combining with 1 of sulphur to form sulphuric acid, what will be the atomic weight of sulphuric acid? Ans. $16+3\times8=40$.

How are the compounds of oxygen with each element below it in Table I. placed? 37. How are the compounds of all the elements below oxygen with each other placed? 37.

What is chloride of sodium composed of? Sulphuret of iron? Sulphuret of hydrogen? Light carburet of hydrogen? Heavy carburet of hydrogen? Ammonia? In ammonia, how many atoms of hydrogen to 1 of nitrogen? Why is NH^3 the symbol for ammonia? Why is 17 the atomic weight of ammonia?

(These symbols show what each compound is made up of. They are not designed to be committed, but to be used for reference. The reader, for instance, might wish to ascertain what carbonic acid *is*. If he turn to this table, he wlll see carbonic acid, CO^2 22. The C shows one atom of carbon, 6; the O^2, two atoms of oxygen $8+8=16$; and so of all the other compounds, and of the salts in Table II. formed from these compounds.)

How many compounds in Table I. are called acids? 39. How many are called oxides? 39. Why are the oxides called also bases? 39. Why are the salts formed from these acids and bases called oxygen salts? 40. How does the name of these salts always end? 40. Are there other salts? 40. If carbonic acid were combined with soda, what would be the name of the salt thus formed? 40. If the soda should take a double portion of the acid, what prefix would precede its name? 40. What other prefix signifies the same as *bi?* 40.

Can you distinguish between those compounds whose name ends in *uret*, and the salts whose names end in *ate?* 41.

What is a *prot*oxide? 42. A *sesqui*oxide? 42. A *per*oxide? 42.

Will you tell me what is the composition of sulphate of iron (copperas)? 43 and Table II. Of sulphate of soda (Glauber's salt)? 43 and Table II. When water exists in crystals, what is it called? 43.

What is the composition of sulphate of lime (plaster, gypsum)?

44 and Table II. How much water is contained in 86 lbs. of plaster? 44. If this be heated to redness, what takes place? 44. Could you now look into Table II., and learn precisely how any of these salts are constituted?

Will you give some account of chloric acid? 45. Of sulphuric acid? 46. Of phosphoric acid? 47. Of carbonic acid? 48. How is carbonic acid constituted? 48 and Table I. What is its form? 48. What its weight? 48 When first formed, what takes place? 48. What takes place soon? 48. What portion of the air on an average is carbonic acid? 48.

Of what do plants consist largely? 48. Whence do they obtain this? 48. How do they receive it? 48. What of the vegetation of the globe? 48.

When vegetable matter is burnt, what becomes of its carbon? 48. When it is eaten? 48. When it decays? 48.

Lime-stone is carbonate of lime; what proportion of it is carbonic acid? 48. What proportion of this is carbon? Table I.

The shells of fish and coral rock are also carbonate of lime; when shells, coral and lime-stone, are burnt into quick-lime, what becomes of the carbonic acid? 48.

What is said of volcanoes? 48. Of some springs? 48. Of fissures in the earth? 48.

What is said of the exhaustion and re-supply of carbonic acid in the air? 48.

Is carbonic acid poisonous to breathe? 48. How much of it is there in pure air? 48. How much in air from the lungs? 48.

Why should school-rooms and churches be often ventilated? 48.

What is silicic acid? 49. By what other name is it more commonly called? 49. How is it composed? 49. In what two states does it exist in soils? 49. What part does silica perform in the growth of crops? 49. What is said of oats grown on peat, in which there is little or no silica? 49. How much of it do we generally find in soils? 49.

Of what is nitric acid composed? 50. What is said of its salts? 50. Of old plastering? 50. Of Chinese gardeners? 50.

What is muriatic acid composed of? 51. What was it formerly called? 51.

What is the composition of water? 52. Give an account of its decomposition and its recomposition? 52.

Where does the protoxide of iron often exist abundantly? 53. Is it hurtful to plants? 53. How may the farmer know whether his land is troubled with it? 53. What is the cure? 53. How do you account for that variegated film that sometimes appears on water? 53. May lime be used in such cases? 53. If ashes are applied, why should the ground be first drained? 53.

How is the sesquioxide of iron composed? 54. How does it differ from the protoxide? 54. What are those scales by the blacksmith's anvil? 54. For what are these good? 54. How are they to be applied? 54.

What gives to some soils their red color? 54. To others their sickly yellow? 54. How can these last be cured? 54.

What can you say of the peroxide, or black oxide of manganese? 55.

How is potash composed? 56. What is said of its caustic power? 56. What has to be combined with potassium, to make it potash? 56. What with potash to make it carbonate of potash? 56. What with that to make it bicarbonate? 56. In what form is it applied to land? 56. In what form does potash exist in ashes? 56. How much carbonate of potash is there in common wood ashes? 56. How much soda? 56. How much lime? 56.

Will you trace sodium through its combinations up to carbonate of soda? 57. To sulphate of soda? 57. What is soda-ash? 57.

What is lime? 58. What is water-slacked lime? How much water does it take in? 58. What is air-slacked lime? 58. Trace the metal calcium through its combinations? 58. What does it form if combined with carbonic acid? 58. With sulphuric acid? 58. With silicic acid? 58. With muriatic acid? 58. What are those substances called which consolidate water in themselves and yet appear to be dry, as slacked lime? 58.

Magnesia is obtained from sea-water and from a species of magnesian lime-stone, called dollomite; it exists in this lime-stone and in sea-water, as carbonate of magnesia; if the carbonic acid is driven off, what does it become? 59.

How is alumina composed? 60. Of what is it the basis? 60. What is pure clay? 60.

What is chloride of sodium? 61. Why may common salt be beneficial to corn, potatoes, and turnips? 61.

How many sulphurets of iron are there? 62. What is the bisulphuret sometimes called? 62. Why? 62.

How is sulphuretted hydrogen composed? 63. It is a light, evanescent gas; where may it often be detected by its smell? 63. What is said of its influence on health? 63. On the growth of plants? 63.

What is the name of that gas which often rises in bubbles in stagnant water? 64. Of that which is used for purposes of lighting? 64. What experiment is mentioned in 64?

What is the composition of ammonia? 65. How is its odor recognized? 65. Where is it generated? 65. If left to its own course, what does it become? 65. Where does it go? 65. How is it brought back to the earth? 65. Can its escape be arrested? 65.

GEOLOGY.

What is the form of the earth? 66. What inference from this

with regard to the state in which it once was? 66. What is its average weight? 66.

How many square miles on the earth's surface? 67. How many square miles of land? 67. How many of water? 67. What is the height of the highest land? 67. The depth of the deepest water? 67. The probable average height of land? 67. The average depth of water? 67.

(If we suppose the population of the globe to be 1000 million, we have 32 acres to each person; if the population should double once in 25 years, there would be, in 200 years, 4 persons to each acre.)

What is said of the crust of the earth? 68. With what is it covered? 68. What is the weight of the atmosphere to the square inch of the earth's surface? 68. To the square foot? 68. Of the whole atmosphere? 68.

What is the difference between stratified and unstratified rocks? 69. How must the unstratified rocks have been formed? 69. What are they called? 69.

How did the stratified rocks receive their present form? 70. What are they called? 70. What else are they called, and why? 70.

Which rocks are the older? 71. Which the newer? 71. Give the illustration? 71.

Could the igneous and the aqueous rocks have been formed at the same time? 72. Why not? 72.

What classes of rocks do we find above the igneous? 73. Give some account of the primary rocks? 74. Of the secondary? 75. Of the tertiary? 76.

What do we find above the tertiary rocks? 77. What is drift? 77. Where is this found? 77. Whence did it come? 77. From how far? 77.

What has been formed above the drift? 78. By what causes? 78.

Which of the formations then is most recent? 79. Which next? 79. Which next? 79. Which next? 79. Which is the lowest of the stratified rocks? 79. On what do these rest? 79.

Are there some portions of igneous rocks above and among the stratified? 80. Whence do they seem to have come? 80.

Have we reason to believe that the earth was created in the form in which it now is? 81. Is there reason to believe that different portions of the earth's crust were formed at periods remote from each other? 82.

From what are all soils formed? 83. Do we know when the drift period was? 83. Describe its action? 83.

What of the loose materials on the earth's surface? 84. Do soils come from the underlaying rock? 84. Whence do they come? 84.

How many simple minerals constitute the mass of known rocks? 85. What are they? 85. What are the binary compounds in rocks? 86.

Will you describe quartz? 87. Felspar? Mica? Hornblende? Carbonate of lime? Talc? Serpentine? 87. What part of the ponderable matter of the globe is oxygen? 89. What part of its crust is silica? 89. What is silica? Table I. How much of its crust is alumina? 89. What is said of potash, or *potassa?* 89. Of soda? 89. Of lime and magnesia? 89. Of iron? 89. Of manganese? 89.

What does Dr. Dana say rocks are? 90. What is quartz? 91. Felspar and mica? 91. Hornblende? 91. Talc and serpentine? 91. What are silicates? Table II. What is said of the quantity of silica in soils?

In which rocks is there more silica? 92. In which more magnesia, alumina and lime? 92. Are rocks a good criterion of soils? 92.

How do soils generally produce when first cultivated? 93. On what does their continuance of fertility depend? 93.

Describe the action of a torrent in depositing its coarser and finer matter? 94. Have other causes done the like on a larger scale? 94. What is the consequence? 94.

Which lands should we cultivate first? 95. Is it probable that poorer lands may pay well hereafter? 96.

What is said of reclaiming lands? 96. By what should farmers be guided? 96. Has science done anything for other employments? 96. What science especially deserves the farmer's attention? 97. Why? 97.

What was thrown up by the most ancient volcanoes? 98. What by those more recent? 98. What by the most recent and by those now in operation? 98. How does the temperature become as we descend into the earth? 98. At what rate does it become warmer? 98. What do we infer from this? 98.

At 40 or 50 miles deep what might we expect to find? 98. What next above the lava? 98. What above the trap? 98. What above the granite? 98. Name all the formations above the granite, beginning with the primary? 98. Do each of these last form an entire layer around the whole earth? 98. Explain the reasons? 98. On what may the cultivable soil lie? If it lies on granite, what is that region called? 98. If on primary rocks? 98. If on secondary? 98. If on tertiary? 98. If on alluvial? 98.

Of what does soil consist? 99. What rock must it have originated from? 99. Whence did all the igneous rocks on and near the earth's surface come? 99. What changes have befallen them from the time of their emission from the earth? 99. What has been mingled with them, to form a soil fit for cultivation, containing the organic as well as the mineral ingredients? 99. Do we know all the agencies by which the Almighty prepared the soil for man? 99. What were some of his agents? 99. Did God make the earth a garden? 99. What did he make it capable of becoming by human agency? 99.

Why have not rural employments been held in the highest honor? 100. What employment is most conducive to rational enjoyment and long life? 100. How did the Creator intend that the farmer should thrive? 100. How has He therefore made his employment? 100.

Chemically considered, what is the difference between good soils and poor? 101. Does a soil which the Creator has perfected by those protracted agencies before spoken of, contain the elements in Table I.? 102. . Is it almost wholly made up of them? 102. Do they exist in it in their elementary state? 102.

Do the binary compounds mentioned in Table I., either exist in soils, or in some way contribute to their fertility? Table I. and 103, 104, &c. What is sulphuric acid? Table I. How much of this might be expected to be found in a good soil? 103. In what state? 103.

What is phosphoric acid? Table I. and 47. How much of this might we expect to find in a good soil? 103. In what state?

What do you say of carbonic acid in soils? 104. Of silicic acid? 105. Of nitric acid? 106. Of water? 107. How does the food of plants enter them? 107. How much oxygen will water absorb or dissolve in itself? 107. How much nitrogen? 107. How much hydrogen? 107. How much carbonic acid? 107. How much ammonia? 107. What does water do with these gases? 107.

What other substances does water dissolve and carry into plants? 107. How does the excess of water then leave the plant? 107. What benefit in irrigating with pure water? 107. What extra benefit in irrigating with impure water? 107. How may such irrigation be considered? 107.

What is said of the oxides of iron in soils? 108. Of the oxides of manganese? 109. Of potash? 110. Of soda? 111. Of lime? 112. Of magnesia? 113. Of alumina? 114. Of chloride of sodium? 115. Of sulphuret of iron? 116. Of sulphuret of hydrogen? 117. Of light carburetted hydrogen? 118. Of ammonia?

Do these binary compounds exist as such in soils? 121. If not, how then? 121.

What do you understand by the inorganic part of a soil? 122. What by the organic part? 122. A stick of oak wood contains about 98 parts of organic matter to two of inorganic; if you burn it, where does the organic part go? Into the air. What becomes of the inorganic part? It falls down as ash. What four elements constitute organic matter? 122. Why are carbon, hydrogen, oxygen and nitrogen, called organic elements? 33.

As vegetable matter decays, does it form organic acids? 122. How many? 122. What are the names? 122. What other organic acids are mentioned? 123. What is the composition of acetic acid (vinegar)? 123. What of oxalic acid (C^2O^3)? 123.

If oxalic acid should combine with potash, soda, lime, or some other base (Table III.), what salts would it form? Oxalate of potash, oxalate of soda, &c. Do all these acids form salts with the bases in a similar way? They do, and they are named from the acid, changing its ending into ate, and the base; as oxalate of lime, acetate of potash, &c., &c.

PLANTS.

What do you say of the well-matured seed? 124. Of the embryo? 125. What further of the embryo? 126. Of what does the germ consist? 127. What is the office of the leaves? 127. Of the roots? 127. Do plants choose their food? Illustration? 127 and 128.

What are the essentials of germination? 129. Whence does the plant derive its first food? 129. When does a plant hate, and when love the light? 130. Will you repeat what are the essentials of germination? 129. When these are supplied, what takes place? 131. Explain this evolution of heat? 131 and 132.

Is acetic acid (vinegar) formed in the seed? 133. For what purpose? What other substance is formed? 133. What power has diastase? 133. Is there sugar in the seed? 133. What is turned into sugar? 133. What takes place in cooking flour? 134. Explain? 135 and 136. During germination, what do seeds absorb, and what emit? 137. What takes place afterwards? 137. Why is this? 137.

What reflection may we make? 138. What suggestion to the husbandman? 138. Will you illustrate this in full? 138. What further is said about *starting* plants well? 139. At whose disposal is a part of what makes plants grow? 140. Whose is another part? If we work our own part rightly, what takes place? 140. What is the moral? 141 and 142.

Do plants purify the air for animals? 143. How? 143. Do animals enrich the air for plants? How? Are they mutually beneficial? 143. Who breathes the best air? 143.

Whence does the plant obtain most of its carbon? 144. Whence the rest? 144. How are its oxygen and hydrogen furnished? 144. How are they taken in? 144. How is the plant furnished with nitrogen? 145. In what do nitric acid and ammonia exist? 145. Are animal manures specially valuable for the nitrogen in them? They are. What does the nitrogen in fermenting animal manures form, if nothing else is present? Volatile ammonia, which escapes and is lost. What does it form, if plenty of peat and a little slacked lime are mixed? In this case the nitrogen forms nitric acid; this combines with the lime, forming nitrate of lime, a most valuable addition to the manure.

Whence does the plant obtain most of its organic elements? 146. How might we say the plant feeds itself? 147. Explain further? 147.

What do you say of the flower-leaves (petals, 148. Does this give them their colors? 148. How? 148.

After flowering, what seems to be their principal effort? 149. Is the growth always a measure of fruitfulness? 149. What is said of manuring corn wholly in the hill? 149. What of corn-roots? 149. Should all the manure then be in one place? 149.

Is late hoeing injurious? 150. Why? 150. Will you explain this fully? 151.

With regard to the circulation of plants, what may be taken as a sort of sample? 153. Of what does the stem consist? 154. What of the pith? 155. Describe the roots? 156. What are the spongioles? 157. Describe the rootlets? 158.

What is said of the branches and twigs? 159. What are the leaf-stems? 160. What flows through them? 160 What does the circulation of the sap through the leaves resemble? 160. Describe the leaves? 160.

When the sap has circulated through the leaves, what takes place? 161. How is the annual layer of wood formed? 161.

What is the destiny of all. that lives? 162. On what do men, brutes and plants live? 162. What of the floating matter around us? 162. What is probable?

What does the plant devour? 163. What happens to it in return? 163.

When plants have passed their maturity, what happens? 164. What are their proximate constituents? 164. What secondary products come from these?

What is said of starch? 165. Of sago? 165. Of arrow-root? 165. Of tapioca? 165. Of all these? 165.

Of what are starch, gum and sugar composed? 166. Of what are gluten, caseine and albumen composed? 167. Why are they called nitrogenous? 167. What do they contain besides the organic elements? 167. Do one or more of them exist in all plants? 168.

What can you say of gluten? 169. Of caseine? 170. Of albumen? 171. How can you separate the constituents of flour? 172 and 173.

Which of the substances just spoken of contain nitrogen? 174. What else? What letters then may characterize them? 174. Which contain no nitrogen? 175. What letters may characterize these? 166. Which are most nutritious, as food? 175.

What remarkable fact is stated of starch, gum and sugar? 175. In what proportions are the oxygen and hydrogen in them? 175. Is the same true of woody fibre? 175. Of what then do they consist?

Are starch, gum and sugar identical in composition? 175 and 176. Of what transformations are they capable? 176.

ANIMALS AND THEIR PRODUCTS.

Besides organic matter, what 12 ingredients enter into soils? 179. Which of these does not pass into plants? 180. Which of the eleven that pass into plants, does not pass into the composition of animals? 180. Through what round do the other ten pass? 180. What is the effect of selling crops? To exhaust the land. What is the effect of selling beef, pork, butter, cheese, &c.? The same, but to a less degree. What would be the effect of selling everything from a farm? 180, end.

What is the prevention? 181. Why may farmers near the city sell all? 181. What is the true way for the great mass of farmers? 181.

What is important for practical farmers? What are the animals to consume mainly the produce of American farmers? 183.

Into what three classes may we divide animals? 184. What return work only for their keeping? 184. What work and growth? 185. What return the products of their bodies only? 186. What is necessary in order that the farmer should get the worth of his feed from animals? 186.

What are the conditions of farming? 187. How must the farmer dispose of his crops? 187. What are his pay-masters? 187. On what condition are they "good pay?" 187. How should he use them? 187. Why? 187. For what other reason? 187. Explain? 187.

What is said of being observant of the habits of animals and attentive to their wants? 188.

What of providing for the comfort of animals both in summer and in winter? 189.

How should animals be supplied with salt? 190. What have some supposed with regard to watering animals? 191. What is the truth in this matter? 191.

What hay should be given to milch cows? 192. What to working cattle and horses? 192. To dry cows? 192. How should young stock be fed? 192.

What two sources does the farmer look to for his remuneration for wintering stock? 193. Will stock cattle pay for their keeping, if fed on good hay only? 193.

Will you illustrate the fact last stated? 194. How must the loss be avoided? 194. What are the equivalents of one lb. of Indian meal mentioned at the close of section 194?

Now although hay alone, given to stock cattle, will not produce an advance in their value equal to its estimated worth, may not a proper mixture of food effect the object? 195. Will you state the argument, as in the 195th section?

Can certain rules be given? 196. What of the feeder? 196. What of feeding? 196.

What is the office of starch, gum and sugar in animal food? 197. Of the nitrogenous substances, gluten, caseine and albumen? 197. Of oil? 197. Of all the organic substances? 197. Of phosphate of lime? 197.

Explain how the non-nitrogenous substances support respiration? 198.

To what may the lungs be compared? 199. How may the starch, gum and sugar be regarded? 199.

Will you give the substance of section 200? Of 201? Of 202? Of 203?

At how many things especially must the farmer look? 204. What is the first? 204. What the second?

What is the farmer to dispose of first? 205. In what condition should his stock face the solid winter? 205. Why? 205. What preparations should he have made? 205.

What is the composition of good meadow hay? 206. What is a most valuable ingredient of the inorganic matter? 206.

What will you say of such hay for the purposes of feeding? 207. How should good early-cut hay be disposed of among the cattle? 207. What of hay for horses? 207.

What is said of the use to be made of less valuable hay? 208. What is a great fault in expending poor hay? 208.

To what account may very poor hay, if the farmer have such, be turned? 209.

Can straw be put to any use, as fodder? 210. What is observable with regard to it? 210.

What is the analysis of Indian corn? 211. In what is the ash peculiarly rich? 211. Why is it very fattening? 211. What is its tendency when given to milch cows? 211.

What is said of corn as food for horses? 212. If horses are fed on corn, should it be old or new? 213. What of corn for fattening sheep? 214. What cheaper food is recommended for store-sheep? 214.

What is the staple for pork-making? 215. Of what opinion are many farmers? 215. Will feeding corn to swine pay in all cases? 215. What should be remembered? 215. What is necessary in order that the making of pork and lard should pay? 216.

Should corn-meal for swine be fermented? 217. There are several degrees of fermentation; what is the first? 217. The second? 217. The third? 217. If Indian meal were passed through all these stages, would it be fattening? 217. If arrested, between the first and second, what is believed? 217.

What is the composition of oats? 218. What is observed with regard to them? 218. What of oat straw? 218. To what should it not be given? 218. Why? 218. What is said of fattening animals? 218.

How does rye compare with corn? 219. How differ? 219. What more is said? 219.

What can you say of growing carrots, and of the use to which they should be put? 220.

What is the composition of turnips, and the best use to be made of them? 221.

What of potatoes? 222. If used for cattle, or other animals, how is their value increased? 222.

What is said of apples raw? 223. Cooked? 223. Of cooking food in general? 223.

Should coarse hay and straw be cut? 224. Why? 224.

What would you say of letting stock to be wintered become poor at the threshold of winter? 225. On the heels of winter? 225. What would you advise with regard to both the fall and the spring? 225. Why? 225.

What advice would you give with regard to young cattle? 226.

To what are all animals subject? 227. Explain this further? 227..

What is said of milk in section 228? In 229? In 230? In 231? In 232? In 233? In 234? In 235? In 236? In 237? In 238?

What is said of butter in 239? In 240? In 241? In 242? In 243? In 244? In 245? In 246? In 247? In 248? In 249? In 250? In 251? In 252? In 253? In 254? In 255?

What is said of cheese in 256? In 257? In 258? In 259? In 260? In 261? In 262? In 263? In 264? In 265? In 266? In 267? In 268? In 269?

MANURES.

Into how many and what classes may lands be distributed with relation to manure? 272.

What three kinds of land belong to the first class? 273. What is said of lands belonging to the second class? 274. On what condition are these to be cultivated? 274. What of lands belonging to the third class? 275.

Of the three soils of which an analysis is given by Professor Johnstone, which exhibits no deficiencies? 276. In what is the second deficient? 276. In what the third?

Would the first of these soils produce any one of the crops mentioned in Table V., without manure? 276 and 277. How does this appear? 276 and 277. How does it appear that the second would produce, by the addition of potash, soda and chlorine? 276 and 277. What of the third? 278.

What would you do with such a soil as the first? 279. As the second? 279 and 280. Would the special manuring, recommended for the second, answer permanently? 281.

What would farming become, if reliable analyses of soils could in all cases be obtained? 282. Explain the benefit of such knowledge? 282.

What of the importa ce of manures? 283. What of good management in this respect? 284.

Will you explain the distinction of manures into animal, vegetable and mineral? 285. What is the difference between manures and stimulants? 286. What are amenders? 286. What is unfortunate for this distinction? 287.

What do you understand by organic matter in soils? 288. How can you ascertain its per cent. in a soil? 289.

What three modes are there of restoring organic matter to soils? 290.

How do the acids exist in soils? 291. Chlorine and soda? 291.

What is said of applying mineral manures in 292?

What have some supposed? 293. If we could know precisely what mineral manures to apply, would these produce fertility permanently? 293. Why not? 293. What would have to be resorted to ere long? 293.

What is the farmer's great resource? 294. What must enrich the farm? 294. How can this be effected? 294.

What is said of the value of manures? 295. What is the golden subject of agriculture? 296.

Into what shape should the surface of the barn-yard be put?

How would you prevent water running downwards into the soil? 299.

Explain the use of peat, swamp mud, &c., as retainers? 300. Whence the great value of these substances for mixing with manures? 301.

What farmers may well purchase fertilizers from abroad? 302. What would you say if those who have not husbanded their home resources, should expend money for fertilizers from abroad? 303.

What is said of making barn-yard manure in section 304? In 305? In 306? What of the value of manure thus composted in the barn-yard? 306.

How many cellars should a barn have? 308. Why should each cool? 308.

What would you place on the bottom of the manure cellar? 309. How much? 309. How should this cellar be constructed? 309. How much composting matter should be in readiness for the winter? 309.

Explain how you would proceed? 310. When will manure so prepared and housed be ready for use? 311. When has too much labor been withheld? 311.

To what use might such manure be put? 312.

In applying such manure, could you exactly meet the wants of the soil? 313. What might you expect if you should supply more of some ingredients than were wanted for the first crop? 313.

What of nitrogen as an ingredient of manures? 314. Wha'

have some thought? 314. Explain the formation of ammonia? 314. Of carbonate of ammonia? 314.

How can the escape of ammonia be prevented? 315. Will you give the explanation in full? 315.

What injury comes from the washing of manures? 316. From excessive fermentation? 316. What example of *burning* manure? 317.

What is said of pig-pen manure in 318? In 319? In 320? In 321? In 322? In 323? In 324?

What should be a rule for manures? 325. How can the washing of manures be prevented during heavy rains? 320 and 325.

In what two conditions will a pig-pen be very offensive? 326. What four bad consequences follow? 326. What then is another rule? 327. How can all the bad consequences before spoken of be prevented? 327. Why should the farmer be more careful than others that no offensive odor arise from his premises? 328. What is a singular but well-known fact? 328.

What is said of the manure of the sheep-fold in section 329? In 330? In 331?

What portable and inoffensive fertilizer is sometimes prepared from night-soil? 332. What advantage arises from this? 332. On a farm, how may night-soil be managed advantageously? 333, 334 and 335.

How may the washings of the sink be best managed and applied?

What is said of composting in 337? In 338? In 339? In 340? In 341? In 342?

Should there be in the vicinity of the house a place of reception for whatever may be of value for the land? 343. Will you describe how it may be managed? 343. Describe further, as in 344? As in 345?

What is said of woollen rags? 346. Of old shoes and boots, and of accumulations of leather parings? 346. Of dead animals? 347. Of bones? 348. What further of bones in 349? In 350?

What would you say of burning bones and then applying the ashes?

What is said of foreign fertilizers? 352. Of the men who undertake to furnish them? 353.

How can the farmer best decide for himself when to go to the expense of purchasing fertilizers from abroad? 354. Till he thus decides, on what must he depend?

What further is said of the importance of home manures in 356? In 357?

Why has the *chemistry of common objects* been dwelt upon in former portions of this work? 358. Why the geological formation of soils? 358. How have plants and animals been spoken of? 358. How manures? 358.

What is said of land in most European countries? 359. How in our own country? 360.

Will you describe the condition of a thriftlessly managed farm? 361. What might the owner have done? 362. What is said of other farms? 363.

What would be the perfection of farming? 364. Explain? 364. Have some farms doubled and some halved the amount of manure? 364. How are the owners? 364.

What does this show? 365. What else does it show? 365.

Is perfection in crop-growing attainab'e? 366. How then must we proceed?

What do you say of the *analysis* and the *examination* of soils? 367. Who only can make reliable analyses? 367. Who can make examinations of soils? 367. What of the observing farmer? 367.

What advice then should be given to the farmer? 368. To what should he be encouraged? 368.

If I were thinking to buy a farm at a *fixed* price, what should I do well, in the first place, to inquire? 370.

If the farm were wholly of one kind of land, what might it be well to do? 371. If there were 8 or 10 varieties? 371.

Will you describe what would be called a *pure clay?* 372, 1. How is a *strong clay soil* constituted? 372, 2. A *clay loam?* 372, 3. A *loam?* 372, 4. A *sandy loam?* 372, 5. A *sandy soil?* 372, 6. *Peat?* 372, 7. *Swamp muck?* 372, 8.

How could you decide for yourself to which of these classes a soil belongs? 373.

How would the existence of all these soils on a farm affect its value? 374. Why not purchase a farm that needs no amendment? 374.

What encouragement do you find for such as are obliged to work farms which need amending? 374. Give an instance where an improvement might be made at a cost less than its probable value? 375.

Will you give another such instance? 376. Another? 377. Another still? 378. In purchasing a farm, what should we look at? 379. What four things should we study? 379.

Will you now repeat how many and what soils, exclusive of peat and swamp muck, we have spoken of? What remains? 380.

What can you say of the density of soils? 381. Will you state what is about the weight of a sandy soil? 382. Of the several other soils here named? 382. What is said of soils retaining heat? 382.

What do you say of the fineness of soils? 383. Of their adhesiveness? 384. Of their power of absorbing moisture? 385. Of their power of containing moisture? 386.

What is capillary attraction? 387. Does this exist in soils? 388. Illustrate this by an experiment? 388. Will you show why water in the subsoil makes a field cold? 389. State the argument in favor of draining wet lands? 390.

When does the water in a soil sink? 391. When does it rise? 391. What do you call that action by which it rises? 387. Will you explain this more fully? 391.

Should a cultivated soil be permeated by the air? 392. What is the pressure of the atmosphere in each square inch of soil? 68. Would this pressure force the air into openings made by the plough and harrow? It would.

What gases does a rich and moist soil take from the air? 392. What possesses this power in a high degree? 393. What inference from this? 393.

What possesses this power of absorbing and holding gases in the next degree? 394. What possesses it in a considerable degree? 394. What in the lowest degree? 394. What argument do we derive from this against entrusting green manures to light soils? 402. What in favor of composting such manures with peat? 402.

What is said of the effect of peat, swamp muck, and fermented manures? 395. Will you describe the effects of mixing clay with sandy soils? 396.

If the farm, before spoken of, has now been bought, what will become the question? 397. State some of the difficulties which the occupant will have to encounter? 397.

What will the farmer find among the first things to be done? 398. Can he apply these manures, so as to give every field exactly what it needs, and no more? 399. How may he apply them? 399.

How may he find things on this farm? 400. How must he take them? 401.

If he should conclude to use half his green manure now, and keep the other half for composting, what would you say of putting the first half into sandy or lightish loamy soils? 402. What objection is there to using it as a top-dressing for mow-land? 402. Might it do well thus? 402. What would be a safer application of it? 402. Why? 402, 393 and 394. What striking instance of loss, by the application of green manure to a sandy soil, is mentioned? 402. How has better corn been raised on similar lands? 402.

What is said of the application of barn-yard manure? 403. What of the composted manure supposed to be found on this farm? 404. What of hog-pen manure? 405. Of sink settlings? 406. Of chip manure? 406.

What is said of the application of night-soil? 407. What does old plastering contain? 408. What is its most valuable ingredient? 408. Is this very soluble? 408. Would you put it on a small space? 364 and 408.

How should our farmer, on his new place, decide whether to

purchase plaster largely or not? 409. Will you give the several effects which plaster produces on soils adapted to it? 410.

Will plaster operate well alone permanently? 411. What does it require? 411. In what two ways is organic matter added when plaster is used on pastures? 411. How should it be added when plaster is used on mow-lands? 411. How on ploughing? 411. Should we complain of plaster because other manuring is required to keep the land permanently good? 411. What of ashes? 411 and 56, near the end.

Does the subject of capillary attraction, as explained in 387 and onward, throw any light on the question of deep ploughing? 413. In what case might it be bad policy to plough deeply? 413. If the soil is deep, and the subsoil compact, what do you say? 414. Reasons? 414.

On all ordinary soils, how deep should we plough at least? 415. If the soil below that depth is impervious to water, what should be done? 415. What two dangers do you thus escape? 415. What positive benefit do you gain? 415.

Explain the operation of water as a carrier of food to plants? 416. Will you illustrate still more fully the necessity of the free passage of water and air through the soil? 417.

What is said of the importance of deep ploughing? 418. What caution is to be observed? 418. What would be the safest course? 419.

What is said of mixing unlike soils? 420. How may this sometimes be done without the labor of transportation? 420. What cause sometimes prevents the good effect of deep ploughing? 420.

How may the farmer judge whether his land is troubled with the protoxide of iron? 420. If it should prove to be so, what may he do?

How many oxides of iron are there in soils? 53 and 54. Which of these is very soluble in water? 53. What is its effect on plants? 53. To what does this oxide turn when exposed to the sun and air? 53.

What is said of a properly prepared soil? 421. How does it sustain vegetation? 421. What auxiliaries has it? 421. What further is said of such a soil? 422.

What is said of the best time for cutting grass? 423. Of the importance of early hoeing? 423. Of the time when wheat, rye and oats should be cut? 423. Which gives most hay, early or late cutting? 423. Which gives the best? 423.

What frequently occupies the attention of farmers after haying? 424. Which of those great improvements, before spoken of, would he be tempted to enter upon first? 424. Why may we suppose that he will prefer to take hold of the business of draining? 424.

What sort of a swamp is it, which he wishes to drain? 424.

How might he proceed? 425. If he hire this work, how should it be done? 426. Why by the job? 426. How could he and the men who should undertake it ascertain what would be a fair remuneration? 426.

How long would the mud thrown up improve in quality by lying? 427. How long at least should it lie? 427. Would it be too heavy to remove when first thrown up? It would.

What further is said of the work to be done on this swamp? 427. What three modes of filling covered drains are mentioned? 428. There would be much labor in reclaiming five acres of such land; would it probably pay? 428.

What is said of brush-drains? 429. Of stone-drains? 429. Of tile-drains? 429. What caution is requisite, that, in laying tile-drains, the ends of the tiles do not get slipped aside from each other and filled up? 429.

What is said of draining lands that are not considered swampy? 430. What name is given to the regular draining of lands, with covered drains, at equal distances from each other? 430, at the end.

If, another season, our farmer should have time and means to attack that ten-acre lot, how might he lay out and prosecute the work? 431 and 432.

The labor of reclaiming and amending lands could hardly "pay" in a new country, and especially if far from market; will you state some reasons for believing it to be a paying business in the Atlantic States, where the produce is near great markets, and where it generally brings a good price? 433.

What would you say of putting clay on sandy soils, if the c'ay lies very near? 434. And what would you say of the prospect of remuneration, provided the land is in the vicinity of a good market? 434.

Which do you think is best paid for his labor, the man who spoils a good farm, or the man who mends a poor one? 434.

If, of two adjacent soils, one was too sandy, and the other too clayey, how would you amend them both? 435. Suppose a peaty and a clayey soil to lie side by side, would the same course be advisable? It would. If a peaty and a sandy soil were very near each other, could you apply the same remedy? I could.

Ordinary soils weigh at about the rate of 1000 tons to the acre, taking them ten inches deep, which is 100 tons for each inch in depth; would it be necessary, therefore, that soils, in order to be mixed with paying results, should lie near each other? It would.

(It may be well enough for you to recollect, that as 1000 tons constitute the whole soil ten inches deep, 100 tons is ten per cent. of the whole; ten tons is one per cent.; and one ton is one-tenth of one per cent.; so that for every ten loads put upon an acre of land, one per cent. is added to the soil, if the ploughing is ten inches deep, or two per cent. if only five.)

What is the prevailing rotation of crops in England called? 436. What is this course? 436. To what soils is it adapted? 436. How is the advantage of this rotation explained? 436. What rotation is there preferred for heavier soils? 436.

What are two important points of difference between English and American agriculture? 437. Does English usage therefore throw much light on our course? 437.

If we look into the analyses of crops, do we see important differences in their requirements? 277, Table V. In the absence of fixed rules, settled, as in European countries, by long practice, what general rule should guide us? 437.

DADD'S (GEO. H.) AMERICAN CATTLE DOCTOR, - - - $1 00

CONTAINING THE NECESSARY INFORMATION FOR PRESERVING THE Health and Curing the Diseases of Oxen, Cows, Sheep, and Swine, with a great variety of Original Recipes and Valuable Information in reference to Farm and Dairy management, whereby every man can be his own Cattle Doctor. The principles taught in this work are, that all Medication shall be subservient to Nature—that all Medicines must be sanative in their operation, and administered with a view of aiding the vital powers, instead of depressing, as heretofore, with the lancet or by poison. By G. H. DADD, M. D., Veterinary Practitioner.

THE DOG AND GUN, - - - - - - - - - - 50

A FEW LOOSE CHAPTERS ON SHOOTING, among which will be found some Anecdotes and Incidents; also, instructions for Dog Breaking, and interesting letters from Sportsmen. By A BAD SHOT.

MORGAN HORSES, - - - - - - - - - - 1 00

A PREMIUM ESSAY ON THE ORIGIN, HISTORY, AND CHARACTERISTICS of this remarkable American Breed of Horses; tracing the Pedigree from the original Justin Morgan, through the most noted of his progeny, down to the present time. With numerous portraits. To which are added hints for Breeding, Breaking, and General Use and Management of Horses, with practical Directions for training them for exhibition at Agricultural Fairs. By D. C. LINSLEY.

SORGHO AND IMPHEE, THE CHINESE AND AFRICAN SUGAR CANES, - - - - - - - - - - 1 00

A COMPLETE TREATISE UPON THEIR ORIGIN AND VARIETIES, CULTURE and Uses, their value as a Forage Crop, and directions for making Sugar, Molasses, Alcohol, Sparkling and Still Wines, Beer, Cider, Vinegar, Paper, Starch, and Dye Stuffs. Fully Illustrated with Drawings of Approved Machinery; With an Appendix by LEONARD WRAY, of Caffraria, and a description of his patented process of crystallizing the juice of the Imphee; with the latest American experiments, including those of 1857, in the South. By HENRY S. OLCOTT. To which are added translations of valuable French Pamphlets, received from the Hon. JOHN Y. MASON, American Minister at Paris.

THE STABLE BOOK, - - - - - - - - - - 1 00

A TREATISE ON THE MANAGEMENT OF HORSES, IN RELATION TO Stabling, Grooming, Feeding, Watering and Working, Construction of Stables, Ventilation, Appendages of Stables, Management of the Feet, and of Diseased and Defective Horses. By JOHN STEWART, Veterinary Surgeon. With Notes and Additions, adapting it to American Food and Climate. By A. B. ALLEN, Editor of the American Agriculturist.

THE HORSE'S FOOT, AND HOW TO KEEP IT SOUND, - 50

WITH CUTS, ILLUSTRATING THE ANATOMY OF THE FOOT, and containing valuable Hints on Shoeing and Stable Management, in Health and in Disease. By WILLIAM MILES.

THE FRUIT GARDEN, - - - - - - - - - 1 25

A TREATISE, INTENDED TO EXPLAIN AND ILLUSTRATE THE PHYSIOLOGY of Fruit Trees, the Theory and Practice of all Operations connected with the Propagation, Transplanting, Pruning and Training of Orchard and Garden Trees, as Standards, Dwarfs, Pyramids, Espalier, &c. The Laying out and Arranging different kinds of Orchards and Gardens, the selection of suitable varieties for different purposes and localities, Gathering and Preserving Fruits, Treatment of Diseases, Destruction of Insects, Description and Uses of Implements, &c. Illustrated with upwards of 150 Figures, representing Different Parts of Trees, all Practical Operations, forms of Trees, Designs for Plantations, Implements, &c. By P. BARRY, of the Mount Hope Nurseries, Rochester, N. Y.

FIELD'S PEAR CULTURE, - - - - - - - - 75

THE PEAR GARDEN; or, a Treatise on the Propagation and Cultivation of the Pear Tree, with Instructions for its Management from the Seedling to the Bearing Tree. By THOMAS W. FIELD.

BRIDGEMAN'S (THOS.) YOUNG GARDENER'S ASSISTANT, $1 50

IN THREE PARTS, Containing Catalogues of Garden and Flower Seed, with Practical Directions under each head for the Cultivation of Culinary Vegetables and Flowers. Also directions for Cultivating Fruit Trees, the Grape Vine, &c., to which is added, a Calendar to each part, showing the work necessary to be done in the various departments each month of the year. One volume octavo.

BRIDGEMAN'S KITCHEN GARDENER'S INSTRUCTOR,	½ Cloth,	50
" " " "	Cloth,	60
BRIDGEMAN'S FLORIST'S GUIDE, - - - - -	½ Cloth,	50
" " " - - -	Cloth,	60
BRIDGEMAN'S FRUIT CULTIVATOR'S MANUAL, -	½ Cloth,	50
" " " " -	Cloth,	60

COLE'S AMERICAN FRUIT BOOK, - - - - - - 50

CONTAINING DIRECTIONS FOR RAISING, PROPAGATING AND MANAGING Fruit Trees, Shrubs and Plants; with a description of the Best Varieties of Fruit, including New and Valuable Kinds.

COLE'S AMERICAN VETERINARIAN, - - - - - 50

CONTAINING DISEASES OF DOMESTIC ANIMALS, THEIR CAUSES, Symptoms and Remedies; with Rules for Restoring and Preserving Health by good management; also for Training and Breeding.

SCHENCK'S GARDENER'S TEXT BOOK, - - - - 50

CONTAINING DIRECTIONS FOR THE FORMATION AND MANAGEMENT of the Kitchen Garden, the Culture and Use of Vegetables, Fruits and Medicinal Herbs.

AMERICAN ARCHITECT, - - - - - - - 6 00

THE AMERICAN ARCHITECT, Comprising original Designs of Cheap Country and Village Residences, with Details, Specifications, Plans and Directions, and an Estimate of the Cost of Each Design. By JOHN W. RITCH, Architect. First and Second Series, 4to, bound in 1 vol.

BUIST'S (ROBERT) AMERICAN FLOWER GARDEN DIRECTORY, 1 25

CONTAINING PRACTICAL DIRECTIONS FOR THE CULTURE OF PLANTS, in the Flower-Garden, Hot-House, Green-House, Rooms or Parlor Windows, for every Month in the Year; with a Description of the Plants most desirable in each, the nature of the Soil and Situation best adapted to their Growth, the Proper Season for Transplanting, &c.; with Instructions for Erecting a Hot-House, Green-House, and Laying out a Flower Garden; the whole adapted to either Large or Small Gardens, with Instructions for Preparing the Soil, Propagating, Planting, Pruning, Training and Fruiting the Grape Vine.

THE AMERICAN BIRD FANCIER, - - - - - -

CONSIDERED WITH REFERENCE TO THE BREEDING, REARING, FEEDing, Management and Peculiarities of Cage and House Birds. Illustrated with Engravings. By D. JAY BROWNE.

REEMELIN'S (CHAS.) VINE DRESSER'S MANUAL, - - 50

AN ILLUSTRATED TREATISE ON VINEYARDS AND WINE-MAKING, containing Full Instructions as to Location and Soil, Preparation of Ground, Selection and Propagation of Vines, the Treatment of Young Vineyards, Trimming and Training the Vines, Manures, and the Making of Wine.

DANA'S MUCK MANUAL, FOR THE USE OF FARMERS, - 1 00

A TREATISE ON THE PHYSICAL AND CHEMICAL PROPERTIES OF Soils and Chemistry of Manures; including, also, the subject of Composts, Artificial Manures and Irrigation. A new edition, with a Chapter on Bones and Superphosphates.

CHEMICAL FIELD LECTURES FOR AGRICULTURISTS, - 1 00

By DR. JULIUS ADOLPHUS STOCKHARDT, Professor in the Royal Academy of Agriculture at Tharant. Translated from the German. Edited, with notes, by JAMES E. TESCHEMACHER.

BUIST'S (ROBERT) FAMILY KITCHEN GARDENER, - - $0 75

Containing Plain and Accurate Descriptions of all the Different Species and Varieties of Culinary Vegetables, with their Botanical, English, French and German names, alphabetically arranged, with the Best Mode of Cultivating them in the Garden or under Glass; also Descriptions and Character of the most Select Fruits, their Management, Propagation, &c. By Robert Buist, author of the "American Flower Garden Directory," &c.

DOMESTIC AND ORNAMENTAL POULTRY. Plain Plates, - 1 00
Do. Do. Do. Colored Plates, - 2 00

A Treatise on the History and Mangement of Ornamental and Domestic Poultry. By Rev. Edmund Saul Dixon, A.M., with large additions by J. J. Kerr, M.D. Illustrated with sixty-five Original Portraits, engraved expressly for this work. Fourth edition revised.

HOW TO BUILD AND VENTILATE HOT-HOUSES, - - 1 25

A Practical Treatise on the Construction, Heating and Ventilation of Hot-Houses, including Conservatories, Green-Houses, Graperies and other kinds of Horticultural Structures, with Practical Directions for their Management, in regard to Light, Heat and Air. Illustrated with numerous engravings. By P. B. Leuchars, Garden Architect.

CHORLTON'S GRAPE-GROWER'S GUIDE, - - - - - 60

Intended Especially for the American Climate. Being a Practical Treatise on the Cultivation of the Grape Vine in each department of Hot House, Cold Grapery, Retarding House and Out-door Culture. With Plans for the Construction of the Requisite Buildings, and giving the best methods for Heating the same. Every department being fully illustrated. By William Chorlton.

NORTON'S (JOHN P.) ELEMENTS OF SCIENTIFIC AGRICULTURE, 60

Or, the Connection between Science and the Art of Practical Farming. Prize Essay of the New York State Agricultural Society. By John P. Norton, M.A., Professor of Scientific Agriculture in Yale College. Adapted to the use of Schools.

JOHNSTON'S (J. F. W.) CATECHISM OF AGRICULTURAL CHEMISTRY AND GEOLOGY, - - - - - - 25

By James F. W. Johnston, M.A., F.R.SS.L. and E., Honorary Member of the Royal Agricultural Society of England, and author of "Lectures on Agricultural Chemistry and Geology." With an Introduction by John Pitkin Norton, M.A., late Professor of Scientific Agriculture in Yale College. With notes and additions by the author, prepared expressly for this edition, and an Appendix compiled by the Superintendent of Education in Nova Scotia. Adapted to the use of Schools.

JOHNSTON'S (J. F. W.) ELEMENTS OF AGRICULTURAL CHEMISTRY AND GEOLOGY, - - - - - - 1 00

With a Complete Analytical and Alphabetical Index and an American Preface. By Hon. Simon Brown, Editor of the "New England Farmer."

JOHNSTON'S (JAMES F. W.) AGRICULTURAL CHEMISTRY, 1 25

Lectures on the Application of Chemistry and Geology to Agriculture. New edition, with an Appendix, containing the Author's Experiments in Practical Agriculture.

THE COMPLETE FARMER AND AMERICAN GARDENER, 1 25

Rural Economist and New American Gardener; Containing a Compendious Epitome of the most Important Branches of Agriculture and Rural Economy; with Practical Directions on the Cultivation of Fruits and Vegetables, including Landscape and Ornamental Gardening. By Thomas G. Fessenden. 2 vols. in one.

FESSENDEN'S (T. G.) AMERICAN KITCHEN GARDENER, - 50

Containing Directions for the Cultivation of Vegetables and Garden Fruits. Cloth.

NASH'S (J. A.) PROGRESSIVE FARMER, - - - - - $0 60

A SCIENTIFIC TREATISE ON AGRICULTURAL CHEMISTRY, THE GEology of Agriculture, on Plants and Animals, Manures and Soils, applied to Practical Agriculture; with a Catechism of Scientific and Practical Agriculture. By J. A. NASH

BRECK'S BOOK OF FLOWERS, - - - - - - 1 00

IN WHICH ARE DESCRIBED ALL THE VARIOUS HARDY HERBACEOUS Perennials, Annuals, Shrubs, Plants and Evergreen Trees, with Directions for their Cultivation.

SMITH'S (C. H. J.) LANDSCAPE GARDENING, PARKS AND PLEASURE GROUNDS, - - - - - - - - - - - 1 25

WITH PRACTICAL NOTES ON COUNTRY RESIDENCES, VILLAS, PUBLIC Parks and Gardens. By CHARLES H. J. SMITH, Landscape Gardener and Garden Architect, &c. With Notes and Additions by LEWIS F. ALLEN, author of "Rural Architecture."

THE COTTON PLANTER'S MANUAL, - - - - - 1 00

BEING A COMPILATION OF FACTS FROM THE BEST AUTHORITIES ON the Culture of Cotton, its Natural History, Chemical Analysis, Trade and Consumption, and embracing a History of Cotton and the Cotton Gin. By J. A. TURNER.

COBBETT'S AMERICAN GARDENER, - - - - - 50

A TREATISE ON THE SITUATION, SOIL, AND LAYING-OUT OF GARDENS, and the making and managing of Hot-Beds and Green-Houses, and on the Propagation and Cultivation of the several sorts of Vegetables, Herbs, Fruits and Flowers.

ALLEN (J. FISK) ON THE CULTURE OF THE GRAPE, - 1 00

A PRACTICAL TREATISE ON THE CULTURE AND TREATMENT OF THE Grape Vine, embracing its History, with Directions for its Treatment in the United States of America, in the Open Air and under Glass Structures, with and without Artificial Heat. By J. FISK ALLEN.

ALLEN'S (R. L.) DISEASES OF DOMESTIC ANIMALS, - 75

BEING A HISTORY AND DESCRIPTION OF THE HORSE, MULE, CATTLE, Sheep, Swine, Poultry, and Farm Dogs, with Directions for their Management, Breeding, Crossing, Rearing, Feeding, and Preparation for a Profitable Market; also, their Diseases and Remedies, together with full Directions for the Management of the Dairy, and the comparative Economy and Advantages of Working Animals, the Horse, Mule, Oxen, &c. By R. L. ALLEN.

ALLEN'S (R. L.) AMERICAN FARM BOOK, - - - - 1 00

THE AMERICAN FARM BOOK; or, a Compend of American Agriculture, being a Practical Treatise on Soils, Manures, Draining, Irrigation, Grasses, Grain, Roots, Fruits, Cotton, Tobacco, Sugar Cane, Rice, and every Staple Product of the United States; with the Best Methods of Planting, Cultivating and Preparation for Market. Illustrated with more than 100 engravings. By R. L. ALLEN.

ALLEN'S (L. F.) RURAL ARCHITECTURE; - - - - 1 25

BEING A COMPLETE DESCRIPTION OF FARM HOUSES, COTTAGES, AND Out Buildings, comprising Wood Houses, Workshops, Tool Houses, Carriage and Wagon Houses, Stables, Smoke and Ash Houses, Ice Houses, Apiaries or Bee Houses, Poultry Houses, Rabbitry, Dovecote, Piggery, Barns, and Sheds for Cattle, &c., &c., together with Lawns, Pleasure Grounds, and Parks; the Flower, Fruit, and Vegetable Garden; also useful and ornamental domestic Animals for the Country Resident, &c., &c. Also, the best method of conducting water into Cattle Yards and Houses. Beautifully illustrated.

WARING'S ELEMENTS OF AGRICULTURE; - - - - 75

A BOOK FOR YOUNG FARMERS, WITH QUESTIONS FOR THE USE OF Schools.

PARDEE (R. G.) ON STRAWBERRY CULTURE; - - $0 60

A Complete Manual for the Cultivation of the Strawberry; with a description of the best varieties.

Also, notices of the Raspberry, Blackberry, Currant, Gooseberry, and Grape; with directions for their cultivation, and the selection of the best varieties. "Every process here recommended has been proved, the plans of others tried, and the result is here given." With a valuable appendix, containing the observations and experience of some of the most successful cultivators of these fruits in our country.

GUENON ON MILCH COWS; - - - - - - - 60

A Treatise on Milch Cows, whereby the Quality and Quantity of Milk which any Cow will give may be accurately determined by observing Natural Marks or External Indications alone; the length of time she will continue to give Milk, &c., &c. By M. Francis Guenon, of Libourne, France. Translated by Nicholas P. Trist, Esq.; with Introduction, Remarks, and Observations on the Cow and the Dairy, by John S. Skinner. Illustrated with numerous engravings. Neatly done up in paper covers, 37 cts.

AMERICAN POULTRY YARD; - - - - - - - 1 00

Comprising the Origin, History and Description of the different Breeds of Domestic Poultry, with complete directions for their Breeding, Crossing, Rearing, Fattening, and Preparation for Market; including specific directions for Caponizing Fowls, and for the Treatment of the Principal Diseases to which they are subject, drawn from authentic sources and personal observation. Illustrated with numerous engravings. By D. J. Browne.

BROWNE'S (D. JAY) FIELD BOOK OF MANURES; - - 1 25

Or, American Muck Book; Treating of the Nature, Properties, Sources, History, and Operations of all the Principal Fertilizers and Manures in Common Use, with specific directions for their Preservation, and Application to the Soil and to Crops; drawn from authentic sources, actual experience, and personal observation, as combined with the Leading Principles of Practical and Scientific Agriculture. By D. Jay Browne.

RANDALL'S (H. S.) SHEEP HUSBANDRY; - - - - 1 25

With an Account of the Different Breeds, and general directions in regard to Summer and Winter Management, Breeding, and the Treatment of Diseases, with Portraits and other Engravings. By Henry S. Randall.

THE SHEPHERD'S OWN BOOK; - - - - - - 2 00

With an Account of the Different Breeds, Diseases and Management of Sheep, and General Directions in regard to Summer and Winter Management, Breeding, and the Treatment of Diseases; with Illustrative Engravings, by Youatt & Randall; embracing Skinner's Notes on the Breed and Management of Sheep in the United States, and on the Culture of Fine Wool.

YOUATT ON SHEEP, - - - - - - - - - - - 75

Their Breed, Management and Diseases, with Illustrative Engravings; to which are added Remarks on the Breeds and Management of Sheep in the United States, and on the Culture of Fine Wool in Silesia. By William Youatt.

YOUATT AND MARTIN ON CATTLE; - - - - - 1 25

Being a Treatise on their Breeds, Management, and Diseases, comprising a full History of the Various Races; their Origin, Breeding, and Merits; their capacity for Beef and Milk. By W. Youatt and W. C. L. Martin. The whole forming a Complete Guide for the Farmer, the Amateur, and the Veterinary Surgeon, with 100 Illustrations. Edited by Ambrose Stevens.

YOUATT ON THE HORSE; - - - - - - - - 1 25

Youatt on the Structure and Diseases of the Horse, with their Remedies. Also, Practical Rules for Buyers, Breeders, Smiths, &c. Edited by W. C. Spooner, M.R.C.V.S. With an account of the Breeds in the United States, by Henry S. Randall.

YOUATT AND MARTIN ON THE HOG; - - - - - $0 75

A Treatise on the Breeds, Management, and Medical Treatment of Swine, with Directions for Salting Pork, and Curing Bacon and Hams. By Wm Youatt, V.S., and W. C. L. Martin. Edited by Ambrose Stevens. Illustrated with Engravings drawn from life

BLAKE'S (REV. JOHN L.) FARMER AT HOME; - 1 25

A Family Text Book for the Country; being a Cyclopedia of Agricultural Implements and Productions, and of the more important topics in Domestic Economy, Science, and Literature, adapted to Rural Life. By Rev. John L. Blake, D D.

MUNN'S (B.) PRACTICAL LAND DRAINER; - - - 50

Being a Treatise on Draining Land, in which the most approved systems of Drainage are explained, and their differences and comparative merits discussed; with full Directions for the Cutting and Making of Drains, with Remarks upon the various materials of which they may be constructed. With many illustrations. By B. Munn, Landscape Gardener.

ELLIOTT'S AMERICAN FRUIT GROWER'S GUIDE IN ORCHARD AND GARDEN; - - - - - - - - - - 1 25

Being a Compend of the History, Modes of Propagation, Culture, &c., of Fruit Trees and Shrubs, with descriptions of nearly all the varieties of Fruits cultivated in this country; and Notes of their adaptation to localities, soils, and a complete list of Fruits worthy of cultivation. By F. R. Elliott, Pomologist.

PRACTICAL FRUIT, FLOWER, AND KITCHEN GARDENER'S COMPANION; - - - - - - - - - 1 00

With a Calendar. By Patrick Neill, LL.D., F.R.S.E., Secretary of the Royal Caledonian Horticultural Society. Adapted to the United States from the fourth edition, revised and improved by the author. Edited by G. Emerson, M D., Editor of "The American Farmer's Encyclopedia." With Notes and Additions by R. G Pardee, author of "Manual of the Strawberry Culture." With illustrations

STEPHENS' (HENRY) BOOK OF THE FARM; - - 4 00

A Complete Guide to the Farmer, Steward, Plowman, Cattleman, Shepherd, Field Worker, and Dairy Maid By Henry Stephens. With Four Hundred and Fifty Illustrations; to which are added Explanatory Notes, Remarks, &c., by J S Skinner. Really one of the best books a farmer can possess

PEDDERS' (JAMES) FARMERS' LAND MEASURER; - - 50

Or, Pocket Companion; Showing at one view the Contents of any Piece of Land from Dimensions taken in Yards. With a set of Useful Agricultural Tables.

WHITE'S (W. N.) GARDENING FOR THE SOUTH; - - 1 25

Or, the Kitchen and Fruit Garden, with the best methods for their Cultivation; together with hints upon Landscape and Flower Gardening; containing modes of culture and descriptions of the species and varieties of the Culinary Vegetables, Fruit Trees, and Fruits, and a select list of Ornamental Trees and Plants, found by trial adapted to the States of the Union south of Pennsylvania, with Gardening Calendars for the same. By Wm. N. White, of Athens, Georgia.

EASTWOOD (B.) ON THE CULTIVATION OF THE CRANBERRY; 50

With a Description of the best Varieties. By B. Eastwood, "Septimus" of the New York Tribune.

AMERICAN BEE-KEEPER'S MANUAL; - - - - - 1 00

Being a Practical Treatise on the History and Domestic Economy of the Honey Bee, embracing a full illustration of the whole subject, with the most approved methods of managing this Insect, through every branch of its Culture; the result of many years' experience. Illustrated with many engravings By T. B. Miner.

THAER'S (ALBERT D.) AGRICULTURE - - - - $2 00

THE PRINCIPLES OF AGRICULTURE, by ALBERT D. THAER; translated by WILLIAM SHAW and CUTHBERT W. JOHNSON, Esq., F.R.S. With a Memoir of the Author. 1 vol. 8vo.

This work is regarded by those who are competent to judge as one of the most beautiful works that has ever appeared on the subject of Agriculture. At the same time that it is eminently practical, it is philosophical, and, even to the general reader, remarkably entertaining.

BOUSSINGAULT'S (J. B.) RURAL ECONOMY, - - 1 25

IN ITS RELATIONS TO CHEMISTRY, PHYSICS, AND METEOROLOGY: or, Chemistry applied to Agriculture. By J. B. BOUSSINGAULT. Translated, with notes, etc., by GEORGE LAW, Agriculturist.

"The work is the fruit of a long life of study and experiment, and its perusal will aid the farmer greatly in obtaining a practical and scientific knowledge of his profession."

MYSTERIES OF BEE-KEEPING EXPLAINED; - - - - 1 00

BEING A COMPLETE ANALYSIS OF THE WHOLE SUBJECT, consisting of the Natural History of Bees; Directions for obtaining the greatest amount of Pure Surplus Honey with the least possible expense; Remedies for losses given, and the Science of Luck fully illustrated; the result of more than twenty years' experience in extensive apiaries. By M. QUINBY.

THE COTTAGE AND FARM BEE-KEEPER; - - - - 50

A PRACTICAL WORK, by a Country Curate.

WEEKS (JOHN M.) ON BEES.—A MANUAL; - - - - 50

OR, AN EASY METHOD OF MANAGING BEES IN THE MOST PROFITABLE manner to their owner; with infallible rules to prevent their destruction by the Moth With an appendix, by WOOSTER A. FLANDERS.

THE ROSE; - - - - - - - - - - - - - 50

BEING A PRACTICAL TREATISE ON THE PROPAGATION, CULTIVATION, and Management of the Rose in all Seasons; with a list of Choice and Approved Varieties, adapted to the Climate of the United States; to which is added full directions for the Treatment of the Dahlia. Illustrated by Engravings.

MOORE'S RURAL HAND BOOKS, - - - - - - - - 1 25

FIRST SERIES, containing Treatises on—

THE HORSE,	THE PESTS OF THE FARM,
THE HOG,	DOMESTIC FOWLS, and
THE HONEY BEE,	THE COW,

SECOND SERIES, containing— - - - - - 1 25

EVERY LADY HER OWN FLOWER GARDENER,	ESSAY ON MANURES,
ELEMENTS OF AGRICULTURE,	AMERICAN KITCHEN GARDENER,
BIRD FANCIER,	AMERICAN ROSE CULTURIST.

THIRD SERIES, containing— - - - - - - 1 25

MILES ON THE HORSE'S FOOT,	VINE DRESSER'S MANUAL,
THE RABBIT FANCIER,	BEE-KEEPER'S CHART,
WEEKS ON BEES,	CHEMISTRY MADE EASY.

FOURTH SERIES, containing— - - - - - 1 25

PERSOZ ON THE VINE,	HOOPER'S DOG AND GUN,
LIEBIG'S FAMILIAR LETTERS,	SKILLFUL HOUSEWIFE,
BROWNE'S MEMOIRS OF INDIAN CORN.	

RICHARDSON ON DOGS: THEIR ORIGIN AND VARIETIES. . 50

DIRECTIONS AS TO THEIR GENERAL MANAGEMENT. With numerous original anecdotes. Also, Complete Instructions as to Treatment under Disease. By H. D. RICHARDSON. Illustrated with numerous wood engravings.

This is not only a cheap work, but one of the best ever published on the Dog.

LIEBIG'S (JUSTUS) FAMILIAR LECTURES ON CHEMISTRY, $0 50

AND ITS RELATION TO COMMERCE, PHYSIOLOGY, AND AGRICULTURE. Edited by JOHN GARDENER, M.D.

BEMENT'S (C. N.) RABBIT FANCIER; - - - - - 50

A TREATISE ON THE BREEDING, REARING, FEEDING, AND GENERAL Management of Rabbits, with remarks upon their diseases and remedies, to which are added full directions for the construction of Hutches, Rabbitries, &c., together with recipes for cooking and dressing for the Table. Beautifully illustrated.

THOMPSON (R. D.) ON THE FOOD OF ANIMALS - - 75

EXPERIMENTAL RESEARCHES ON THE FOOD OF ANIMALS AND THE Fattening of Cattle; with remarks on the Food of Man. Based upon Experiments undertaken by order of the British Government, by ROBERT DUNDAS THOMPSON, M.D., Lecturer on Practical Chemistry, University of Glasgow.

THE WESTERN FRUIT BOOK; - - - - - - 1 25

BEING A COMPEND OF THE HISTORY, MODES OF PROPAGATION, CULture, &c., of Fruit Trees and Shrubs, &c., &c. By F. R. ELLIOTT.

THE SKILLFUL HOUSEWIFE; - - - - - - - - 50

OR COMPLETE GUIDE TO DOMESTIC COCKERY. TASTE, COMFORT, AND Economy, embracing 659 recipes pertaining to Household Duties, the care of Health. Gardening, Birds, Education of Children, &c., &c. By Mrs L. G. ABELL.

THE AMERICAN FLORIST'S GUIDE; - - - - - - 75

COMPRISING THE AMERICAN ROSE CULTURIST AND EVERY LADY HER own Flower Gardener.

EVERY LADY HER OWN FLOWER GARDENER; - - 50

ADDRESSED TO THE INDUSTRIOUS AND ECONOMICAL ONLY; containing simple and practical Directions for Cultivating Plants and Flowers; also, Hints for the Management of Flowers in Rooms, with brief Botanical Descriptions of Plants and Flowers. The whole in plain and simple language. By LOUISA JOHNSON.

FISH CULTURE; - - - - - - - - - - - 1 00

A TREATISE ON THE ARTIFICIAL PROPAGATION OF CERTAIN KINDS OF Fish, with the description and habits of such kinds as are most suitable for pisciculture. Also directions for the most successful methods of Angling, illustrated with numerous engravings By THEODATUS GARLICK, M.D., Vice President of Cleveland Academy of Natural Science.

FLINT ON GRASSES; - - - - - - - - - - 1 25

A PRACTICAL TREATISE ON GRASSES AND FORAGE PLANTS, COMPRISing their natural history, comparative nutritive value, methods of cultivating, cutting, and curing, and the management of grass lands. By CHAS. L. FLINT, A.M., Secretary of Mass. State Board of Agriculture.

WARDER ON HEDGES AND EVERGREENS; - - - - 1 00

A MANUAL ON LIVE FENCES, WITH PARTICULAR DIRECTIONS FOR THEIR planting, culture and trimming, especially with regard to the Maclura hedges, and how to make it. Also an essay on Evergreens, their varieties, propagation, transplanting and culture in the United States. By JOHN A. WARDER, M.D., President of Cincinnati Horticultural Society.

MOORE'S

Hand Books of Rural and Domestic Economy.

All arranged and adapted to the Use of American Farmers.

PRICE 25 CENTS EACH.

HOGS;

THEIR ORIGIN, VARIETIES AND MANAGEMENT, with a View to Profit, and Treatment under Disease; also Plain Directions relative to the most approved modes of preserving their Flesh. By H. D. RICHARDSON, author of "The Hive and the Honey Bee," &c., &c. With illustrations—12mo.

THE HIVE AND THE HONEY BEE;

WITH PLAIN DIRECTIONS FOR OBTAINING A CONSIDERABLE ANNUAL Income from this branch of Rural Economy; also an Account of the Diseases of Bees and their Remedies, and Remarks as to their Enemies, and the best mode of protecting the Hives from their attacks. By H. D RICHARDSON. With illustrations.

DOMESTIC FOWLS;

THEIR NATURAL HISTORY, BREEDING, REARING, AND GENERAL Management. By H. D. RICHARDSON, author of "The Natural History of the Fossil Deer," &c. With illustrations.

THE HORSE;

THEIR ORIGIN AND VARIETIES; WITH PLAIN DIRECTIONS AS TO THE Breeding, Rearing, and General Management, with Instructions as to the Treatment of Disease. Handsomely illustrated—12mo. By H. D. RICHARDSON.

THE ROSE;

THE AMERICAN ROSE CULTURIST; being a Practical Treatise on the Propagation, Cultivation, and Management in all Seasons, &c. With full directions for the Treatment of the Dahlia.

THE PESTS OF THE FARM;

WITH INSTRUCIONS FOR THEIR EXTIRPATION; being a Manual of Plain Directions for the certain Destruction of every description of Vermin. With numerous illustrations on Wood.

AN ESSAY ON MANURES;

SUBMITTED TO THE TRUSTEES OF THE MASSACHUSETTS SOCIETY FOR Promoting Agriculture, for their Premium. By SAMUEL H. DANA.

THE AMERICAN BIRD FANCIER;

CONSIDERED WITH REFERENCE TO THE BREEDING, REARING, FEEDing, Management, and Peculiarities of Cage and House Birds. Illustrated with Engravings. By D. JAY BROWNE.

CHEMISTRY MADE EASY;

FOR THE USE OF FARMERS. By J. TOPHAM.

ELEMENTS OF AGRICULTURE;

TRANSLATED FROM THE FRENCH, and Adapted to the use of American Farmers. By F. G. SKINNER.

www.ingramcontent.com/pod-product-compliance
Lightning Source LLC
LaVergne TN
LVHW010243110826
845151LV00004B/1380